GRAVITY

GRAVITY

From Falling Apples to Supermassive Black Holes

Second edition

Nicholas Mee

OXFORD
UNIVERSITY PRESS

OXFORD
UNIVERSITY PRESS

Great Clarendon Street, Oxford, OX2 6DP,
United Kingdom

Oxford University Press is a department of the University of Oxford.
It furthers the University's objective of excellence in research, scholarship,
and education by publishing worldwide. Oxford is a registered trade mark of
Oxford University Press in the UK and in certain other countries

First edition published by Virtual Image Publishing in 2014 © Nicholas Mee
Second edition published by Oxford University Press in 2022 © Nicholas Mee

Impression: 2

Published in the United States of America by Oxford University Press
198 Madison Avenue, New York, NY 10016, United States of America

British Library Cataloguing in Publication Data

Data available

Library of Congress Control Number: 2022931921

ISBN 978–0–19–284528–3

DOI: 10.1093/oso/9780192845283.001.0001

Printed in the UK by
Bell & Bain Ltd., Glasgow
Cover image: NASA/CXC/M.Weiss – Public domain

Dedicated to

John Frank Mee

(1934–2021)

Acknowledgments

Many thanks to Debra Nightingale and Philippa Morris for reading draft copies of the manuscript and for their valuable comments and suggestions. Thank you to all my friends and family for their encouragement in what has been a very difficult time.

Thank you to everyone who has joined my mailing list at: www.quantumwavepublishing.co.uk. I am very grateful for all the feedback on my blog articles on the Quantum Wave website. Similarly, thank you to everyone who has subscribed to my YouTube channel at: www.youtube.com/The_Cosmic_Mystery_Tour.

Thank you to Sonke Adlung, Giulia Lipparini, and everyone at Oxford University Press for turning my manuscript into such a beautiful book.

Contents

Can You Feel the Force?

I aim to convince you that you have never felt the force of gravity—strange, but true.

Newton's analysis of gravity provided the vital impetus that kick-started the modern scientific age, yet Newton's theory is still widely misunderstood. School pupils are often taught that there is no gravity in space, and that is why astronauts feel weightless. This misleading statement is passed on even though the main motivation for Newton's theory was to explain the motion of the planets through space around the Sun.

Clearly, it is absurd to suggest that there is no gravity in space. The Earth and Moon are bound together by their mutual gravitational attraction, and the whole solar system is held together by gravity. Indeed, the hundreds of billions of stars that straddle the night sky as the Milky Way are whirling around in a gravitational embrace. Although gravity diminishes with distance, it never disappears completely. Our galaxy, which has a diameter of 100,000 light years, is held together by the gravitational attraction of vast numbers of stars and the other matter that it contains.

So why do astronauts feel weightless while the rest of us, with our feet planted firmly on the ground, are weighed down by more than our responsibilities? The first thing to note is that gravity affects all objects in the same way. We know when an object is acted on by a force because, in accordance with Newton's definition, its velocity changes. In other words, it accelerates. The force of gravity is special because it produces the same acceleration for all objects. This means that atoms, electrons, mobile phones, elephants, spaceships, and planets all fall in the same way under the force of gravity.

We can appreciate how this makes gravity special by considering a different force. Take the electromagnetic force. Electromagnetism affects different materials in different ways. A positive electric charge will repel another positive charge and will attract a negative charge, but it will not affect an uncharged object. But the degree to which a charged object is affected by this repulsion or attraction depends on its mass; the greater its mass, the more difficult it is to change its motion. We express this by saying that a more massive object has more inertia. This means that the electrostatic force generates a greater acceleration for a lightweight charged object than for a more massive object with the same charge.

What is different about gravity? According to Newton's theory, the strength of the gravitational attraction between two bodies depends on their masses. Mass plays a similar role in gravity to that played by electric charge in electromagnetism. So, the gravitational force that an object feels is determined by its mass—the greater the mass, the greater the force. However, as already mentioned, the greater the mass, the greater the inertia, so the greater the mass of an object, the more difficult it is to change its rate of motion. It is a unique feature of gravity that mass plays this dual role. An increase in mass increases the size of the force, and simultaneously decreases the size of the response to the force. The result is a perfect cancellation of these two effects. Therefore the acceleration of an object due to the force of gravity is independent of its mass. In other words, all objects undergo the same acceleration due to gravity, whatever their mass. This was first demonstrated by Galileo over 400 years ago.

The result is that the trajectory followed by an object, when acted on solely by the force of gravity, is the same irrespective of its mass. Earth-dwellers often find this quite hard to accept, which is why the Apollo astronauts offered an explicit demonstration while on the Moon. After the completion of a moonwalk, the commander of Apollo 15, David Scott, held out a geological hammer and a feather and dropped them simultaneously in front of a video camera (Figure 0.1).[1] In the absence of any atmosphere, the hammer and feather fell without any air resistance; the only force acting upon them was the Moon's gravity and, sure enough, both fell with the same acceleration and reached the Moon's surface together.[2]

What would it feel like to fall towards the Moon's surface along with the feather and the hammer? We would fall at the same rate as these other objects and hit the ground alongside both. Even more importantly, all the parts of our body would be accelerated in exactly the same way. Our head would fall with the same acceleration as our kneecaps. Our feet would fall with the same acceleration as our boots, consequently we would not feel anything at all—we would be weightless. There is nothing special about gravity on the Moon. What is special about the Moon is that it is airless and, as there is no air resistance, the only force acting on us is gravity.

Figure 0.1 Apollo 15 astronaut David Scott is standing on the Moon's surface holding a hammer in his right hand and a feather in his left.

If we found ourselves in orbit—on a visit to the International Space Station, perhaps—we would again be in a situation where the only force acting on us is gravity (Figure 0.2). Our spacecraft and our selves within it would be accelerated towards the Earth, but our orbital motion would ensure that we continue to loop the Earth. We would again feel weightless, because every part of our body—every atom—would be simultaneously undergoing the same acceleration.

To see why this would make us feel weightless, we can do a quick *thought experiment*. Rather than performing a real experiment, we can just invent a hypothetical scenario and deduce what its physical consequences would be. Imagine that gravity affects all

Figure 0.2 Astronaut Bruce McCandless II floats in Earth orbit during the first untethered spacewalk on 7 February 1984, as part of NASA's Space Shuttle mission STS-41-B.

materials in the same way except gold and, in our orbiting space-craft, we are wearing a gold ring. According to our hypothetical scenario, the acceleration of the gold is different from the acceleration of our body and, most specifically, of our finger, so we can feel the gold ring tugging on the finger as gravity attempts to pull it away from us. Our finger has to exert a force on the ring to resist this pull, or the ring will be ripped from our finger. What we feel is the force produced by the finger that prevents the ring from escaping. If gold really were affected by gravity in a different way from all other substances, then we would feel nothing apart from this tug on our finger (we might also lose our gold fillings). In short, if the only force acting on us is gravity, such as when we are in orbit, then we feel no force at all—we feel weightless.

Now we need to plant our feet firmly back on the ground. If, as I am asserting, we cannot feel the force of gravity, what is it that we feel when we are on the Earth, as we usually are? When we stand on the ground, gravity is not the only force that is acting on us. We can be sure of this because forces produce accelerations, so, if only a single force were acting on us, we would be accelerating, and if this force were gravity, we would be accelerating towards the centre of the Earth. In fact, there is another force acting on us that exactly balances the force of gravity—the force that prevents us from falling through the floor. This is what we feel. (If a hole opened up beneath us, there would no longer be an upward force resisting gravity and we would accelerate downwards. Now the only force acting on us would be gravity and we would feel weightless.)

It might seem strange that the ground is always conveniently able to produce an upwards force that exactly balances gravity but we can see why if we consider it in the right way. When we stand on the ground, we compress the material that we are standing on, and this pushes the atoms that it is composed of slightly closer together. The outer electrons in the atoms resist being pushed together and repel each other due to their electric charge. This is, ultimately, the origin of the upwards force on us, and it is electromagnetic in origin. The electromagnetic force is so much stronger than gravity that many materials, such as metals or stone, are barely affected by the downward pressure of our weight. Other materials, such as blancmange, where the intermolecular bonding is much weaker, are significantly distorted.

This is not the whole of the story, however. We do not just feel the upward force in our feet or whatever part of our body is in contact with the ground. If we had the consistency of jellyfish, then our bodies would spread into a thin film, with all parts in contact with the ground. Fortunately, we have bones, muscles, tendons, and ligaments that enable us to resist the pull of gravity. These components give our body structure and transmit forces throughout our bodies. When we stand still, the forces are balanced at all points within us, and there is no overall force on any part of our body. In order to balance the forces, there is tension in our muscles and tendons. Our nerves can detect this tension, and this is what we feel as our weight. But all these forces within our body are ultimately electromagnetic in origin, being due to the interactions between the electrons and protons in the proteins and other molecules that compose our body.

Spaghetti Bolognese

I have argued that you have never felt the force of gravity, and this is true. Nevertheless, it is still possible, in principle, to feel gravity in certain extreme circumstances. Gravity diminishes with distance. Newton demonstrated that gravity obeys an inverse square law, which means that if we double the distance between two objects, the gravitational attraction between them falls to a quarter of its original value; if we treble the distance, the force is reduced to a ninth of its original value, and so on. This means that as we stand on the Earth, the gravitational force on our feet is slightly greater than the gravitational force on our head, simply because our feet are about 2 metres closer to the centre of the Earth. However, the Earth's gravity is so feeble that we will never notice this effect.

On the other hand, if we were to venture too close to a very dense object with an extremely intense gravitational field, such as a black hole, our predicament would be rather different. The parts of our body that were closer to the black hole would feel a stronger gravitational force than those that are further away, simply because gravity diminishes with distance. If we were to fall feet-first towards a black hole, for instance, then our feet would be almost two metres closer to the black hole than our head. Near the black hole, its gravitational attraction increases significantly even over a distance as short as a couple of metres. This means that the force pulling our feet towards the black hole is greater than the force pulling our head towards the black hole. If our feet are not to be accelerated away from our head, then our muscles and tendons must take the strain and exert a force to keep our body together. Of course, this is a battle that we cannot win. As we fall towards the black hole, we will be stretched like spaghetti, eventually producing a very messy Bolognese before falling into the black hole.

Fortunately, it is very unlikely that anybody will ever find themselves close enough to a black hole to perform this experiment. Nonetheless, this stretching action of gravity has a dramatic effect here on the Earth. It produces the tides, as Newton first realized. For this reason, it is known as tidal gravity.

This book is the story of gravity and the heroic efforts to make sense of this mysterious feature of all our lives. We will take a look at Sir Isaac Newton's theory of gravity and its publication in his masterpiece, the *Principia*, the book that launched the modern scientific age. Newton's theory ruled for over 200 years until it was superseded by a very different theory based on the curvature of space and time. Albert Einstein was the author of this revolutionary theory. One mind-bending result of Einstein's theory is that there are regions of space that operate like one-way trapdoors from which nothing can escape, not even light. These objects are known as black holes. We will investigate their properties and the ideas of Stephen Hawking, who showed that they might not be totally black after all. But do such outlandish objects really exist? The evidence for black holes is now overwhelming. Physicists routinely detect the thunderous roars of these mighty beasts as they hurtle together and merge and, as we will see, we even have images of the blazing hot plasma swirling around supermassive black holes that are billions of times the mass of the Sun. First we must travel back in time and take a look at the origins of astronomy.

1

The Cosmic Puzzle

He is glorified not in one, but in countless suns, not in a single earth, a single world,
but in a thousand thousand, I say in an infinity of worlds.
GIORDANO BRUNO, *On the Infinite Universe and Worlds (1584)*
'Introductory Epistle'

The Universe Set in Stone

Westminster Abbey is the last resting place of Sir Isaac Newton, the greatest scientific figure of any age, who transformed our understanding of the universe. If we walk past Newton's tomb towards the High Altar, we find that the desire for a compact and unified description of the entire cosmos did not begin with Newton. In front of the High Altar is the Cosmati Pavement[1] (Figure 1.1), which dates from the thirteenth century and the reign of Henry III. It is a remarkable representation of the medieval cosmos in the heart of London. The pavement is constructed from an array of stones and tiles set in an orderly but complex design, formed of squares and circles bound together by ribbons of stone. The pavement represents a medieval vision of the universe. The Cosmati Pavement includes four large roundels set around a square. At the heart of the design, within the square, is a quincunx formed of four smaller roundels with a large fifth roundel at the very centre. Each of the four inner roundels contains a different design: one includes a circle, one a hexagon, one a heptagon, and one an octagon. The central roundel holds a large, beautiful circular piece of chaotically veined marble.

Figure 1.1 The Cosmati Pavement in Westminster Abbey.

The full meaning of the Cosmati Pavement is unknown. There are clues, however, that offer us an insight into its design and meaning. Originally, the ribbons of stone connecting the large roundels carried a message in metal lettering. Only a few of the letters remain, but some of their indentations can still be seen, and there are historical records giving the full, cryptic Latin text. Around the innermost roundel, the message originally said:

Here is the perfectly rounded sphere which reveals the eternal pattern of the universe.

So, this innermost text tells us that the pavement represents the fundamental structure of the cosmos.

Around the border, the text once said:

In the year of Christ one thousand two hundred and twelve plus sixty minus four, King Henry III, the City, Odoricus and the Abbot set in place these porphyry stones.

This is an unusual numerological record of the date of the pavement's construction. The year 1268 has been converted into $1212 + 60 - 4$. This may be because the numbers twelve, sixty, and four were held to have a cosmological significance due to their relationship to familiar time periods; there are twelve months in a year, sixty minutes in an hour, and four seasons of the year.

The longest message was around the four inner roundels, where it once said:

> *If the reader thoughtfully reflects upon all that is laid down, he will discover here the measure of the primum mobile: the hedge stands for three years, add in turn dogs, and horses and men, stags and ravens, eagles, huge whales, the world: each that follows triples the years of the one before.*

This inscription gives a measure of the time span of the universe. It can be reconstructed as:

> *A hedge lives 3 years.*
> *A dog lives 9 years.*
> *A horse lives 27 years.*
> *A man lives 81 years.*
> *A stag lives 243 years.*
> *A raven lives 729 years.*
> *An eagle lives 2,187 years.*
> *A whale lives 6,561 years.*
> *The universe will last for 19,683 years.*[2]

In short, the time span of the universe will be 3^9 years (three to the power of nine, or three to the power of three squared). The doctrine of the Trinity gives the number three a profound significance in Christian theology. No doubt this is part of the numerological mystery of the puzzle.[3] Until recently, the Cosmati Pavement was only uncovered for important state occasions, such as the coronation of the monarch. However, following a major conservation programme (Figure 1.2), it is now on permanent view to the public.

Figure 1.2 The Cosmati Pavement during conservation work.

The Crystal Spheres

Much of the Cosmati design derives from the ideas of the Greek philosopher Aristotle, who was one of the great figures of antiquity. Aristotle lived in the fourth century BC. He was a highly original and penetrating thinker with encyclopaedic interests, who established the bedrock for many of our philosophical and scientific disciplines. Aristotle aimed for nothing less than a complete understanding of the entire physical universe. Aristotle's ideas about physics and astronomy are contained in two of his works that are known as the *Physica* (*On Nature*) and *De Caelo* (*On the Heavens*). Aristotle's physics was developed from a few simple notions based on his intuitions about how the universe works.

Aristotle held the common-sense view that the Earth is positioned at rest at the centre of the universe. Five planets have been known since ancient times: Mercury, Venus, Mars, Jupiter, and Saturn. Each is easily visible to the naked eye, and each can be clearly distinguished from the stars that form the background landscape of the night sky. While the relative positions of the stars remain fixed from night to night, the position of each planet gradually changes. The planets remain within the band of stars that we know as the zodiac, but each follows its own trajectory through these background stars. This is the origin of the word *planet*, which means *wanderer* in Greek. In ancient times, the planetary system was believed to reach as far as Saturn and no further.

Aristotle proposed that each planet occupies its own sphere surrounding the Earth. He believed that all substances in the sub-lunar region below the sphere of the Moon were composed of various combinations of four elements: Earth, Air, Fire, and Water.[4] This was the region of change and transformation, corruption, and decay. The heavens, which stretched outwards from the sphere of the Moon, were composed of a fifth transcendental element known as Aether. The four terrestrial elements were partially separated into layers according to their density. Earth formed the innermost layer, which was surrounded by a layer of Water, composing the world's oceans. Surrounding this, the atmosphere formed a layer of Air and, beyond that, according to Aristotle, was a layer of Fire. The earthbound elements were assumed to have completely different properties from those of the fifth and cosmic element. The next feature of Aristotle's great scheme was the idea of *natural* motion. He believed that in the vicinity of the Earth, the natural motion of the elements was in straight lines: downwards for Earth, upwards for Fire, and horizontally for Air and Water. And when an object reached its natural place, it would stop moving. Any other movement of an object would be forced or *unnatural* motion which could only occur if it were pushed by an animate force such as a human hand, and when the pushing ceased, the object would grind to a halt.

This may have seemed perfectly reasonable to Aristotle and his contemporaries, but it is one of the most serious misconceptions in the history of physics. It is based on everyday life in an environment where friction and air resistance are almost inescapable.

Aristotle knew of ships and horse-drawn carts, but he had never travelled in a car, so he did not experience the feeling of being thrown forward when the brakes are applied sharply. It is simply not true that a force is required to maintain movement. The truth is almost the converse: our motion remains constant and in a straight line, unless we are acted on by a force. In the modern world we are familiar with many modes of transport, and we have all experienced inertia. We know what it feels like when an aeroplane accelerates along a runway, and we know what the effect will be on our bodies when the car we are in changes direction as it turns a sharp bend. Understanding inertia was one of the most important steps in the birth of modern science.

Aristotle believed the heavens were composed of a celestial material that behaved very differently to terrestrial matter. He claimed that the natural motion of this aetherial substance was circular, and that, due to its natural circular motion, there could be no change in the eternal cycling of the heavens. Aristotle believed the heavens were perfect and incorruptible, whereas all terrestrial matter, because of its tendency towards linear motion, was subject to dissolution and decay. In order to fit in with his scheme, Aristotle assumed that any changeable or temporary features of the skies, such as comets, must be produced in the atmosphere; hence the connection between meteors and our name for the study of weather—*meteorology*. The only trace of what we regard as gravity in this scheme was the tendency for heavy objects to fall towards their natural resting place at the centre of the universe, which Aristotle identified with the Earth. Aristotle believed that heavier objects fall towards the Earth faster than light objects.

Motion in the heavens had no counterpart on Earth, so there was no sense in which the movement of the planets was controlled by gravity. According to Aristotle, the planets were held within crystal spheres that revolve and guide the planets around the Earth. Beyond the sphere of the Moon were the spheres of Mercury, Venus, the Sun, Mars, Jupiter, and Saturn; ordered to match the rate at which the planets move across the sky amidst the background stars. Beyond Saturn was the sphere of the fixed stars, and beyond that lay the outermost sphere—the *primum mobile*—the first moved. This was the source of the motion of all nine spheres of the heavens. And because, in Aristotle's view, nothing could move without the action of a mover, it was necessary that it should be turned by the hand of God—the prime mover. The spheres were packed together sufficiently closely for the motion of the outer ones to be transmitted inwards to keep the whole cosmos in motion.

Aristotle constructed his physics from the ground upwards on rather shaky foundations. His ideas about forces and motion and the relationship between them do not genuinely represent the way the universe works. Even so, the brilliance of Aristotle's deliberations dazzled and misled Western thinkers for thousands of years. Aristotle had constructed an intricate and coherent tower of reasoning that offered a very appealing vision of the universe that could provide a consistent answer to virtually any question that might be posed. This fortress of logic would prove almost impregnable for later thinkers.

Unfortunately, it was all utter nonsense. The entire model was constructed by supposedly watertight rational arguments, but the principles on which it was built were derived from Aristotle's personal intuition, and not from experiment, which effectively meant that the entire edifice was built on sand. In the words of the twentieth-century philosopher Bertrand Russell: 'hardly a sentence in either [of Aristotle's books on physics] can be accepted in the light of modern science'.[5] Even so, Aristotle's ideas enchanted Western thinkers for many centuries.

The Divine Comedy

Aristotle's scientific works survived in Arabic translations from their original Greek. Following their translation from Arabic into

Latin in the twelfth century, his extremely coherent but totally erroneous view of the universe was adopted by the Church and elaborated with all the trappings of medieval Christianity. As in Aristotle's original system, the Earth was considered to be stationary at the centre of the cosmos, surrounded by nine concentric crystal spheres. The heavens were supposed to be inhabited by a ninefold hierarchy of angels, whose function was to turn the celestial spheres at just the required rate. In mirror image fashion, the Earth was imagined to contain nine concentric levels of Hell with Lucifer's throne at the centre. The universe that emerged from this synthesis was like a great gothic cathedral. It was filled with beautiful imagery, but it was also covered in hideous gargoyles. Nonetheless, it holds an undeniable aesthetic appeal.[6] It was one of the most unified visions of reality ever devised.[7]

The medieval universe found its ultimate expression in the epic poetry of Dante Alighieri's sublime *Divine Comedy* (Figure 1.3), completed in 1321. In the poem, Dante travels through the cosmos, visiting every corner of the Christian universe. His journey begins in the underworld—the Inferno. Guided by the Roman poet Virgil, Dante travels down through the nine concentric rings that constitute the pit of Hell and witnesses the torments of every category of Earthly sinner until he reaches the realm of Satan himself. After traversing the depths of Hell, Dante emerges to climb the nine tiers of Mount Purgatory before ascending into the Heavens, where he passes through each of the nine celestial spheres.

The *Divine Comedy* is suffused with symbolism and mysticism and brims over with numerological references. The overall structure of the poem is based around the numbers three and nine. The poem is divided into three parts, corresponding to the three regions of the cosmos—Inferno, Purgatory, and Paradise. Each of these parts is further divided into thirty-three sections or cantos which along with the first introductory canto, makes a total of 100. The geography of each of the three regions is organized into

Figure 1.3 Mural of Dante painted in 1465 by Domenico di Michelino in Florence Cathedral. The sinners pass downwards into Hell. In the background is the nine-tiered Mount Purgatory and in the sky are the nine crystal spheres of the heavens. The cathedral itself is shown in the middle ground.

nine concentric realms—the nine circles of Hell, the nine tiers of Mount Purgatory, and the nine celestial spheres of Paradise. The incidents described in the poem also occur in threes. Even the sins and the sinners are organized in threes. There are three rivers of Hell and three mouths of Satan. The poetry itself is written in an intricate rhyming scheme devised by Dante and known as *terza rima*, which interlocks the lines of the poem in threes. This can be expressed symbolically as (aba, bcb, cdc . . .), where a, b, c, etc. represent the rhyming sounds at the end of successive lines.[8] After exploring the labyrinthine geometry of the cosmos, Dante finally enters the Empyrium and the presence of God. In the following verse, He likens his inability to comprehend the meaning of this overwhelming experience fully to the struggle of a geometer faced with an insoluble problem.

As the geometer his mind applies
To square the circle, not for all his wit
Finds the right formula, howe'er he tries
So strove I with that wonder—how to fit
The image to the sphere; so sought to see
How it maintained the point of rest in it.[9]

A Sage Who Knew His Onions

Even the earliest humans must have been aware of the motion of the Sun, Moon, planets, and stars. Astronomy is the oldest science: indeed, it was the contemplation of the skies that ultimately led to the birth of science. There are two intertwined threads that lead from ancient Greece to the astronomy of medieval Europe. One was spun by philosophers[10] such as Aristotle and rested on simple physical and geometrical principles. The other was teased out by professional astronomers who, over the course of many ages, developed the first precise observational science. The main occupation of the astronomers was to record the positions of the Sun, Moon, and planets and to predict their future positions. The purpose of their celestial vigilance was to divine the religious and astrological implications of the planetary wanderings, and this was combined with other vital activities, such as time-keeping. Understanding the cycles of the Sun and Moon was critical for tracking the passage of time and maintaining a reliable and well-tuned calendar. There are other longer cyclical patterns in the heavens. The path of the planet Venus follows an eight-year cycle through our skies,[11] for example, and eclipses of the Sun repeat in eighteen-year cycles known as *saros* cycles. More precisely, the length of a saros cycle is 18 years, $11\frac{1}{3}$ days. There was a total eclipse on 18 March 1988, for instance, that was visible in the Philippines and Indonesia. There was another total eclipse 18 years, $11\frac{1}{3}$ days later and one-third of the way around the globe (due the extra one-third rotation of the Earth); I was lucky enough to see this in Turkey on 29 March 2006. In April 2024, the Sun will be eclipsed again. This time the eclipse will be visible in North America. These three eclipses form part of a series known to modern astronomers as saros cycle 139.[12] The solar eclipses that occur between these eclipses belong to other saros cycles.[13] The Greek astronomers were greatly indebted to their Babylonian predecessors who had compiled observations of the Sun, Moon, and planets stretching back into the dim and distant past. The Babylonians were aware of the cycles of the heavens and could use them to predict the future positions of the celestial bodies. The earliest surviving record of a specific astronomical prediction is attributed to the mathematician Thales of Miletus, who was regarded as one of the Seven Sages of Greece. According to the historian Herodotus,[14] during a five-year war between the Lydians and the Medes, battle was engaged at a site in modern-day Turkey when suddenly day was turned to night. Thales had foretold this event and, when the spectacle duly arrived, it was enough to induce the warring sides to make peace.[15] Since ancient times, this incident has been interpreted as the prediction of an eclipse of the Sun. Modern historians believe that this was the eclipse that took place on 28 May 585 BC.[16] It is assumed that Thales could make this prediction because he was aware of eclipse cycles discovered by the Babylonians and had access to their eclipse records.[17]

The Babylonians may have been skilled observers but they saw the planets as gods and, as far as we know, they did not seek a physical mechanism to account for the planetary motions. Although they discerned periodic patterns in the motion of the heavenly bodies from their long-term records, they apparently felt no need to enquire into the origin of these cycles. The quest for a mechanical or geometrical explanation of the cosmos began with the analytical approach of Greek mathematicians such as Thales, and this was a very important first step towards modern science. The earliest such models that we know of date to a couple of centuries later and are closely related to Aristotle's model that we considered earlier where the stationary Earth is surrounded by celestial spheres that carry the planets, and the heavens in their entirety travel around the Earth once each day. Superimposed on the daily rotation of the heavens, the Sun took its year-long course around the Earth and this accounted for the annual cycle of the seasons.

The task of constructing a system which could accurately describe the paths of the planets across the sky was a good deal more challenging. Guided by philosophers such as Aristotle, astronomers adopted the rule that the circle was the perfect geometrical figure, and the only one that was suitable to represent the motions of the planets. This meant that their attention was restricted to systems in which the heavenly bodies follow circular paths. A second rule insisted that the planets must follow these circles at a uniform speed. If the planets really did cross the sky at a uniform speed, then predicting their motion would be simple, but this is not how the planetary system works. Indeed, there are times when a planet reverses its course and loops backwards before travelling onwards again. This periodic *retrograde* motion, as it is known, makes the heavens much more difficult to understand than Aristotle's simple scheme would suggest.

Apollonius of Perga, a mathematician of the third century BC, proposed a solution that met the demands of circular motion. His idea was that the planets did not simply orbit the Earth on circular paths, as this did not match what was seen in the skies, rather, each planet traced out a circle, known as an *epicycle*, whose centre simultaneously moved around the Earth on a circular orbit (Figure 1.4). Thus, the basic principle of circular motion could be preserved, but the planet itself, as seen from Earth, would follow a more complicated orbit. This epicyclic system of circles on circles remained a feature of every detailed astronomical model for 2000 years.

Apollonius of Perga is also famous for a geometry textbook, known as *The Conics*, which analyses the shapes that are produced

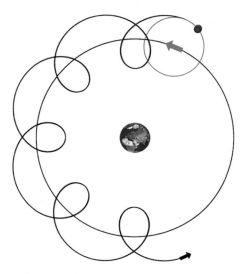

Figure 1.4 In the epicyclic systems of Hipparchus and Ptolemy each planet moved on a circle, known as an epicycle, around an imaginary point that moved around the Earth on a larger circle, known as the deferent.

by slicing a cone at different angles. These shapes are circles, ellipses, parabolas, and hyperbolas. It is a remarkable fact that this very book contains the seed that would eventually germinate and blossom in the mind of an astronomer of the distant future, who would overthrow the age-long obsession with epicycles, as we will see in Chapter 2.

The Antikythera Mechanism

The greatest astronomer of antiquity lived 400 years after Thales, from around 190 BC to around 120 BC. His name was Hipparchus, and he came from Nicaea, which is now the city of Iznik, in Turkey. Like Thales, he is thought to have based his astronomy on the observations and techniques of the Babylonians. Nonetheless, Hipparchus was a very original researcher. He invented many astronomical instruments, such as the astrolabe, and he catalogued the positions of almost 1000 of the brightest stars. Hipparchus was also a great mathematician and is credited with the invention of trigonometry. He is known to have constructed a detailed system of the heavens, based on epicycles that gave a comparatively simple geometrical method to predict the positions of each of the planets in the night sky.

In the year 1900, a shipwreck was discovered by sponge divers off the coast of the Greek island of Antikythera. For over 2000 years, the wreck had been holding a mysterious artefact that sheds light on the technology of the age of Hipparchus and the ancient Greek conception of the heavens. The divers found a green, corroded lump of metal containing a collection of toothed wheels (Figure 1.5).

This remarkable object is now housed in the National Archaeological Museum in Athens. Archaeologists and astronomers have pondered its purpose for over a century. Recent X-ray analyses,

Figure 1.5 Fragment A of the Antikythera mechanism.

along with a great deal of painstaking detective work, have finally revealed its secrets. Within the corroded fragments are the remains of twenty-seven gear wheels and originally there would have been more. The Antikythera mechanism, as it is known, is believed to be the remains of a device that could calculate the positions of the Sun, Moon, and planets, and even predict eclipses by replicating the saros cycle with its gears.[18] From the number of teeth on each wheel it is possible to deduce which cosmic cycle it corresponds to. For instance, one of the wheels has 223 teeth, which is the number of lunar months in a saros cycle. Although the teeth are badly corroded, fragments of text on the device have been deciphered using sophisticated photographic enhancement techniques, and the text confirms that this was indeed the original number of teeth. A reconstruction of the Antikythera mechanism is shown in Figure 1.6.

No-one knows who made the Antikythera mechanism but it gives a unique glimpse of ancient Greek technology. The intermeshed gears of the mechanism are a concrete realization of the ideas of astronomers such as Hipparchus in metal. The apparatus

Figure 1.6 Exploded Reconstruction of the Antikythera mechanism.
Credit: © 2020 Tony Freeth, Images First Ltd.

is so complex and sophisticated that it cannot have been unique; it must have been part of a long tradition of such devices. One day, perhaps, other examples of such machines from the distant past will be discovered.

Rewinding the Epicyclic Clock

The system of epicycles functions rather well. The reason for this, from a modern perspective, is that the main circle, or deferent, represents the planet's orbit around the Sun, while the epicycle represents the Earth's orbit around the Sun. By combining these two circles the centre of the system is shifted from the Sun to the Earth, which gives a reasonable approximation of the path followed by the planet, as seen from Earth. This works because the orbits of the Earth and planets are close to circular. We do not know the details of the earliest systems, but Hipparchus may have included additional circles to further increase its accuracy. The entire planetary system would have kept track of the movement of the planets for many years, but eventually, the synchronization with the heavens would be lost. At this point astronomers of a later generation would need to make new measurements of the planets and bring the system back into synchronization, effectively rewinding the epicyclic clock. One of these later astronomers was Claudius Ptolemy (c.90 AD–c.168 AD) who lived three centuries after Hipparchus. It was Ptolemy who passed on the astronomy of the ancient world to the future.

Ptolemy was a Greek who lived in the leading intellectual centre of his age, the Egyptian city of Alexandria, which was at that time part of the Roman Empire. Ptolemy relied heavily on his predecessor Hipparchus, but he made his own important improvements to the epicyclic system. It was the Ptolemaic version of the universe that was known to later Arabic and European astronomers. Ptolemy's astronomical system was contained in one of four books known to the ancients as the *Tetrabiblos*, the other books being concerned with geography, astrology, and philosophy.

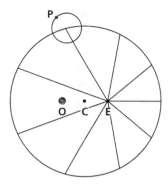

Figure 1.7 The diagram shows a deferent and its epicycle. The labels are: O—the Earth, C—the centre of the deferent, E—the equant point, P—a planet.

Ptolemy's book of astronomy was translated into Latin from Arabic intermediaries and was known to medieval Europe as the *Almagest* (The Greatest). Ptolemy's universe was the last in a long tradition of ancient cosmologies which took as their starting point the idea that all motion of the heavenly bodies must be circular. The full epicyclic system was very complicated and included various devices to improve its accuracy. One significant feature of Ptolemy's system was that the centre of each planet's deferent was located near to, but not actually in the Earth. At an equal distance from the centre, on the opposite side to the Earth, was the *equant point*. As viewed from this point, the centre of the planet's epicycle would appear to revolve around the deferent at a steady rate, as shown in Figure 1.7. This might seem an odd way to comply with the philosophical demand for regular circular motion. Yet, it does fit the observed motion of the planets rather well. It shows the sophistication of the epicyclic systems and the lengths that were taken to follow the philosophical directives while at the same time producing a precise description of the planetary movements.

As viewed from the equant point, the vacant point at the centre of the epicycle moves around the deferent at a constant rate. In the diagram, the deferent is divided into nine arcs. The point at the centre of the epicycle would pass through these nine arcs in equal periods of time.

Although Ptolemy's epicycles offered a computational procedure by which the motion of the planets could be accurately predicted, the overall apparatus was extremely complicated and cumbersome. It was an exercise in geometry designed to *save the appearances*, which meant that it could predict the positions of the planets but it offered no physical mechanism to explain how or why the planets should move in such a way. The system was purely descriptive. Nevertheless, it would not be superseded until the sixteenth century.

The Copernican Revolution

Copernicus is the Latinized name of the Polish cleric and astronomer Mikołaj Kopernik (1473–1543). Copernicus was educated first at the University of Krakow and then in Italy at the University of Bologna. During his student days in Bologna in the 1490s, Copernicus began to question the validity of the traditional and ancient Ptolemaic system of the planets. He believed its fundamental flaw was its *geocentrism*; the planting of the Earth stationary at the centre of the universe. Copernicus argued that placing the Sun at the centre would work much better. In this new *heliocentric* model, the Earth and all the planets would orbit the Sun and the Earth would rotate on its axis once every day. Sometime around 1514, Copernicus wrote an outline of his ideas in a tract that is known as the *Commentariolus*. Its full title in English is *A Commentary on the Theories of the Motions of Heavenly Objects from Their Arrangements*. This was circulated in manuscript form among Europe's leading intellectuals, but Copernicus was hesitant about publishing a fully developed model to rival Ptolemy's.

The big advantage of the heliocentric system was that the stars were fixed in place. Their apparent daily rotation would be due to the rotation of the Earth and not a frantic whirling motion of the entire heavens. As Copernicus well knew, he was not the first to propose such an idea; the Greek astronomer Aristarchus had devised a similar model in antiquity. Copernicus felt that this added strength to his argument. There was, however, a very good reason why the heliocentric model had failed to catch on. If the Earth orbits the Sun once every year, then there should be a seasonal shift in the positions of the stars. It was the failure to observe such shifts that convinced the ancient astronomers that the Earth must reside at the centre of the universe.

The only response that Copernicus could offer was that the stars must be very distant—much more distant than anyone had previously considered possible. This is what made Copernicus' proposal so explosive. It would shatter the cosy medieval world view. Prior to Copernicus, the Church could represent the universe as a relatively small, enclosed space where everything had its purpose, and which had been created for the benefit of mankind. The problem was not so much that the Earth moved rather than the Sun but the implication that, if Copernicus was correct, the universe must be vast. The heliocentric model opened up the possibility that the universe might be essentially infinite in scale. In effect, this was the spatial equivalent of Darwin's temporal expansion of the universe three centuries later. Darwin's ideas about evolution by natural selection implied that the Earth must be ancient; it could not have been created as the dwelling place for humans in the recent past as the Church claimed. Copernicus' proposal had similar implications.

So why *did* people believe that the Earth was stationary at the centre of the Universe, when it is so obvious to everyone today that the Earth is orbiting the Sun?

A Journey to the Stars

Most ancient astronomers believed that the Earth lay at the centre of the cosmos. The strongest argument for this belief was that the relative positions of the stars were fixed during the course of a year. If the Earth orbits the Sun, then the positions of the stars should shift as the Earth moves from one side of the Sun to the other. This effect is known as parallax.

We can get an immediate experience of parallax if we close one eye then open it and close the other. Nearby objects seem to shift in relation to more distant objects, as our viewpoint changes by a few centimetres to the left or to the right. Similarly, as the Earth's position changes by the enormous distance from one side of the Sun to the other, we would expect the apparent positions of nearby stars to shift relative to the more distant stars.

Ancient astronomers observed no seasonal change in the shape of the constellations, so they believed that the Earth was stationary. They had no idea how distant the stars were but assumed that they were probably just beyond the planets because, according to Aristotle, it was the rotation of the sphere of the fixed stars that drove the spheres of the planets. This implied that the stars lay just beyond the orbit of Saturn.

When Copernicus placed the Sun at the centre and overthrew the ancient Earth-centred cosmology, he offered no new evidence to support this radical rearrangement of the heavens. Astronomers still had not detected any shift in star positions due to parallax. But if Copernicus' proposal was accepted, then the only possible explanation was that the stars lay at truly vast distances.[19]

The Distance to the Stars

To work out the distance to a star from a parallax measurement, an astronomer would measure its position at one time in the year—say, 1 January—and then measure its position six months later on 1 July from a viewpoint on the other side of the Earth's orbit (Figure 1.8). The shift in the star's position, coupled with some simple geometry, will then reveal the star's distance. If the shift in position were one degree, then the distance to the star would be

Figure 1.8 Diagram showing the much exaggerated shift in position of a celestial body due to parallax. When the Earth is on one side of its orbit, the body will appear in a different position against the background stars relative to its position when viewed from the other side of the Earth's orbit.

about sixty times the diameter of the Earth's orbit.[20] And a smaller shift implies a greater distance.

In the days of Copernicus, the distance to the Sun was un-known. We now know that the diameter of the Earth's orbit is about 300 million kilometres. This means that a star whose position shifted by as much as one degree during the course of the year would be 60 × 300 million kilometres = 18,000 million kilometres distant. This is about four times the distance to Neptune.

The Moon's apparent diameter is around half a degree, so a shift in position as large as one degree over the course of a year would be very noticeable. Shifts due to parallax are seen in the positions of the planets, and these shifts are the reason why the outer planets move in retrograde loops when the Earth overtakes them each year.[21] This means that even the nearest stars must be much more distant than the outer planets.

It turns out that the shifts in the nearest stars are less than one thousandth of a degree. They are measured in seconds of arc, or *arcseconds*, as they are known. There are 60 minutes of arc in one degree, and 60 seconds in one minute, so the distance to a star that shifts by just one second of arc is 3600 times the distance of an object with a shift of one degree. 3600 × 18,000 million is approximately 65 trillion (where trillion means 1 million million), so such a star would be around 65 trillion kilometres distant—a distance of around six and a half light years. (Light travels 300,000 kilometres in a second and just under 10 trillion kilometres in a year. The light year is therefore a convenient unit with which to express the distances to the stars).

It was not until the nineteenth century that accurate telescopic observations first enabled astronomers to measure the tiny shift in the position of a star due to the Earth orbiting the Sun. The astronomer who first achieved this was the German Friedrich Wilhelm Bessel, who announced in 1838 that he had measured the parallactic shift of the nearby star 61 Cygni. Bessel measured the shift to be about two-thirds of an arcsecond,[22] and so calculated that 61 Cygni must be 10.4 light years away.[23] 61 Cygni is in our cosmic backyard; it is one of our nearest stellar neighbours. Its incredible distance explains why the constellations retain their shape throughout the year, and why the shift due to parallax was undetectable in antiquity.[24]

In 1989, the European Space Agency (ESA) launched the satellite Hipparcos, named after the greatest of the ancient Greek astronomers who, as mentioned, drew up an early star catalogue. The satellite's name also represents a rather contrived acronym 'HIgh Precision PARallax COllecting Satellite' and, as the acronym suggests, one of its main goals was to measure the shift in the apparent positions of stars due to parallax. During its four-year mission, Hipparcos compiled a high-precision catalogue of 100,000 stars, measuring their positions in the sky to within 1 milli-arcsecond. This enabled astronomers to determine the distances to around 20,000 stars with greater than 10% accuracy.

In December 2013, ESA launched a spacecraft named Gaia to build on the work of Hipparcos. Gaia is an incredibly ambitious project that is amassing data on the positions and motion of around 1 billion stars with unprecedented accuracy. It is capable of measuring a star's parallax to within 20 micro-arcseconds. This information is being used to calculate precise distances to a significant proportion of the stars in the Milky Way. Distances to around 20 million stars will be measured to within 1% accuracy and those to a further 200 million stars will be measured to within 10% accuracy. This data has a multitude of astrophysical applications; it will help to refine our models of the stars, as well as our understanding of the size of the universe. Gaia's data will also be used to create a three-dimensional picture of our galaxy in order to better understand its history and evolution. The Gaia mission will continue until sometime around 2024.

An Infinity of Worlds

Copernicus completed the manuscript describing his heliocentric system in around 1532, but he was very hesitant about releasing it to the world. Despite the encouragement of his friends and even some high-ranking Church officials, it was not until he lay on his deathbed in 1543 that his book *De revolutionibus orbium coelestium* (On the Revolutions of the Celestial Spheres) was finally published. The influence of *De Revolutionibus*, as it is usually known, was not due to the details of Copernicus' model and how they compared to the ancient geocentric models. Its success derived from the fact that it contains a simple idea, a *sound bite*, perhaps, that could be easily disseminated—*the Earth moves around the Sun*. The power of this statement is clear. Very few people study celestial mechanics, but everyone knows that the Earth orbits the Sun.

Copernicus' caution was not unwarranted. The full implications of his model were spelt out towards the end of the century by the Italian Dominican friar Giordano Bruno (1548–1600), who drew a series of logical conclusions from Copernicus. Bruno believed that the universe was vast, possibly even infinite in extent, that the Sun was simply one star among many and that other stars must have planetary systems similar to our own. He concluded that some were probably home to other civilizations. In 1584, Bruno wrote:

> *Thus is the excellence of God magnified and the greatness of his kingdom made manifest; he is glorified not in one, but in countless suns; not in a single earth, a single world, but in a thousand thousand, I say in an infinity of worlds.*
>
> GIORDANO BRUNO,
> *On the Infinite Universe and Worlds* (1584)[25]

Bruno travelled Europe, expounding these views, and published pamphlets in Venice and England where he expressed them forcefully. On his return to Venice in 1592, he was denounced to the authorities, who arrested him and then passed him on to the Roman Inquisition. He was charged with holding opinions contrary to the Catholic faith, including claiming the existence of a plurality of worlds and their eternity. After a long trial, on 17 February 1600 Bruno was tortured, stripped naked, and paraded through the streets of Rome. He was then suspended upside down and burnt at the stake as a heretic in the marketplace, the Campo de' Fiori, where his statue now stands (Figure 1.9).

Copernicus did away with the Earth-centred cosmology of the ancients, but he did not see the great opportunity that this offered for simplifying the planetary system. The epicycle that a planet supposedly follows is actually a projection onto the sky of the

Figure 1.9 Statue of Giordano Bruno by Ettore Ferrari in the Campo de' Fiori, Rome.

Earth's motion around the Sun, so placing the Sun at the centre of the system removes the need for epicycles. Yet Copernicus retained almost all the features of the ancient systems, including equant points and epicycles. The result was that his model was a rickety and convoluted mish-mash of ideas. It *saved appearances*, just as the older models had, and the mechanical circles on circles could be used to compute the future positions of the planets to a reasonable accuracy, but this was done in a manner that was incredibly complicated and inelegant. Nonetheless, the idea of placing the Sun at the centre gave a whole new perspective on the mechanics of the solar system. It would now be possible to seek a more physical description of how the planets move.

This advance in our understanding of the universe would be the life work of Johannes Kepler (1571–1630). The heliocentric model was Kepler's starting point. He would not be content with a model of the solar system that simply gave reasonable predictions of the planetary positions. Kepler aimed to know the mind of God, to discover the fundamental blueprint of the cosmos. And it was not a general plan that Kepler sought but the precise details of the planetary system.

2

The Secret of the Universe

There are more things in heaven and earth, Horatio,
Than are dreamt of in your philosophy.
WILLIAM SHAKESPEARE, *Hamlet* (1603), ACT I SCENE 5

What Is Your Star Sign?

Just as the roots of chemistry lie in alchemy, there is no denying that the roots of astronomy lie in astrology. The traditional astrology of the Middle Ages was much more geometrical than its bastardized descendent, the newspaper horoscope. The medieval astrologer required the precise time and date of an event such as his patron's birth. He would then calculate the positions of the planets and the angles between them before considering their implications for his client.

The planetary orbits all lie in a plane that cuts through the Sun at its equator; when projected onto the sky, it is called the ecliptic. The paths of the planets across the sky always lie near to this circle. Long ago the ancient Babylonians organized the prominent stars close to this circle into a convenient collection of twelve constellations. This was useful because, during a year, the Sun spends one month in each constellation. In fact, these constellations are not all the same size, so astrologers neatly spliced the ecliptic into twelve equal 30° chunks and labelled each with the name of the nearest constellation. We know these twelve signpost constellations as the signs of the zodiac. Figure 2.1 is a beautiful early fifteenth-century depiction of the zodiac from a book of hours showing supposed celestial influences on the

Figure 2.1 The Anatomical Man from the Très Riches Heures du Duc de Berry created by the Limbourg Brothers.

human anatomy. The fact that the planets are confined to the vicinity of the ecliptic means it is natural to plot an astrological chart on a disc. The astrologer labels the circle surrounding his disc with the twelve signs of the zodiac and then marks the position of each planet around the circle.

The most important feature of the planetary positions for a nativity horoscope was the angle that was formed between one planet, the soul of the newly born infant on Earth, and a second planet. The astrologers believed that these angles, known as aspects, would determine the fate of the individual. At certain angles there would be a cosmic resonance that would heighten the influence of each planet. For instance, if the two planets were in the same part of the zodiac, so that the angle was zero, they would be in conjunction which greatly increased their influence. Other important angles were 60°, which was known as sextiles; 90°, known as quadrature; 120°, or trines; and 180°, or opposition—in the case of the Sun and Moon this would be at the time of Full Moon.

Even the great astronomer Johannes Kepler (1571–1630) relied on astrology to supplement his income (Figure 2.2). He described it as the *handmaiden of astronomy*. Along with most of his contemporaries, Kepler believed that the Sun, the Moon, and the planets each have a significant astrological influence on life on Earth. It just was not clear exactly what that influence was. But he set out to find it; in his words:

> *one must separate the precious stones from the dung, one must glorify the honour of God, by taking for one's purpose the contemplation of nature, must lift up others by one's own example and exert oneself to move into the bright daylight from the darkness of the human race.*[1]

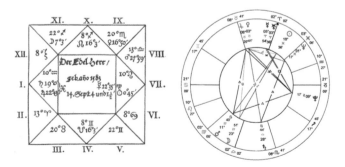

Figure 2.2 Left: Horoscope of General Albrecht von Wallenstein drawn up Kepler in 1608. Right: Circular horoscope depicting the angles between the planets, known as aspects.

The geometrical calculations that Kepler would use to transform astronomy into a modern science would be a close cousin of the astrological geometry of aspects.

Kepler had an unruly upbringing in the small German town of Weil-der-Stadt. His father was a drunken rogue who abandoned his family to fight as a mercenary for the Catholic forces in the Low Countries even though he was a Lutheran himself. Kepler's mother was a herbalist who concocted potions and meddled in other people's affairs, making many enemies. Kepler's childhood was spent with his mother and siblings, living in a tavern owned by his grandparents. His interest in astronomy developed at an early age. He recalled much later that his mother had taken him to view an eclipse of the Moon and, on another occasion, the comet of 1577. Despite his disorderly family life, Kepler gained a good education and his intelligence impressed everyone he met. After school, he went to the University of Tubingen to study theology, with the intention of entering the priesthood. At university, his tutor Michael Maestlin introduced him to the Sun-centred system of the planets devised by Copernicus. Maestlin presented the Copernican system as a useful mathematical tool for making calculations without committing himself to its validity. But Kepler knew at once that it must be correct.

Cosmic Symmetry

After university, Kepler was employed in a seminary school in the Austrian city of Graz. Kepler was an enthusiastic teacher of mathematics and astronomy, but his ideas came thick and fast, so his lessons were filled with endless digressions as his mind wandered among the stars, while his pupils were left behind in the classroom.

Kepler spent his time between lessons pondering the rules devised by God to construct the system of the universe. He was certain that there must be a simple relationship that would explain why there were only six planets and why they were arranged in just the way that we observe. During one lesson in

July 1595, he had a revelation that would inspire his life's work. Kepler drew a diagram for his pupils illustrating how successive conjunctions between the two outer planets Jupiter and Saturn move around the zodiac. Roughly speaking, Jupiter takes twelve years to orbit the Sun and Saturn takes around thirty years.[2] So, in twenty years Saturn has completed two-thirds of an orbit and Jupiter has completed one full orbit and two-thirds of a second. This means that twenty years after one conjunction, the two planets meet again in the night sky, two-thirds of the way around the zodiac. Thus, conjunctions between Jupiter and Saturn occur every thirty years, and three successive conjunctions are located at the corners of a zodiacal triangle, or nearly so, as can be seen in Figure 2.3.

Figure 2.3 Kepler's diagram showing the positions of successive conjunctions of Jupiter and Saturn around the zodiac between 1583, when the conjunction occurred in the constellation Pisces and 1763, by which time the conjunction would have moved a few degrees further round the zodiac into the constellation of Aries.

After drawing the sequence of conjunctions shown in Figure 2.4, Kepler was struck by the circle at the centre of the diagram formed by the edges of the lines between successive conjunctions. It looked like the orbit of Jupiter was inscribed within the orbit of Saturn, such that if the outer edge of the diagram represented Saturn's orbit, then the inner circle might represent Jupiter's orbit. In a flash, Kepler had an idea that seemed to explain the sizes of the planetary orbits in terms of geometry. If he was right, he had discovered part of God's master plan for the universe.

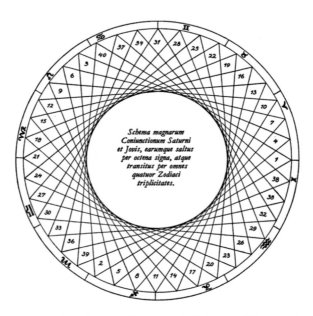

Figure 2.4 Kepler's diagram showing a full circuit of 40 successive conjunctions over a period of 800 years. Conjunction number 1 is shown on the boundary between Pisces and Aries, which is the start of the astronomical year (corresponding to the spring equinox), and very close to the actual position of the conjunction in 1583—the one immediately preceding Kepler's speculations.

The idea was this: starting with a circle to represent Saturn's orbit, he could draw an equilateral triangle within the circle—just like in the diagram—and then, within this triangle, he could draw a circle, such that the circle would touch the midpoint of each edge of the triangle. This second circle would then represent the orbit of Jupiter, and its size would be completely determined by the geometrical construction. Kepler's next step was to draw a square within Jupiter's orbit and within the square a circle that would represent the orbit of Mars. The sequence of polygons and circles could then be continued to determine the orbits of the other planets. Within the orbit of Mars would be a regular pentagon, and the circle within the pentagon would represent the orbit of the Earth. Next would be a hexagon whose inner circle would be the orbit of Venus. Within this orbit would be a heptagon and finally within the heptagon would be the orbit of Mercury (Figure 2.5).[3]

In the middle of his lesson, Kepler was lost in this reverie on the structure of the cosmos. He does not record the reaction of the pupils to his flight of fancy. No doubt, by this point, they had lost all interest in the ramblings of their eccentric teacher.

> **Puzzle 1** A collection of polygons that covers a plane without leaving any gaps is known as a tessellation. The tessellation is regular if the polygons are all regular and of the same type. For instance, there is a regular tessellation formed from equilateral triangles, with six of the triangles meeting at each vertex. There is also a tessellation of squares, where four squares meet at each vertex, and a tessellation of regular hexagons, where three hexagons meet at each vertex. Why is it only possible to construct these three regular tessellations?
>
> HINT: The angles of an equilateral triangle are $60°$, the angles of a square are $90°$, and the angles of a regular hexagon are $120°$.

Over the following months, Kepler sought out information on the dimensions of the planetary orbits to compare his idea to

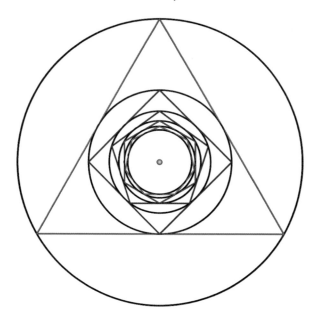

Figure 2.5 Kepler's Polygonal Model. The outer circle represents the orbit of Saturn. Moving inwards the other circles represent the orbits of Jupiter, Mars, Earth, Venus, and Mercury, with the Sun at the centre.

the best astronomical observations. Although the absolute sizes of the planetary orbits were unknown in Kepler's time, the relative sizes of the orbits were well known. Unfortunately, Kepler's model didn't fit at all. He tried altering the order of the polygons but he couldn't find a solution that accurately accounted for the distances between the planetary orbits.

That might have been the end of the matter and no-one would ever have known about this curious idea. A few months later, however, Kepler had a new idea. Instead of fitting regular polygons between the circular orbits of the planets, he would inscribe regular polyhedra between the crystal spheres. Since antiquity it had been known that there are just five regular polyhedra, also known as the Platonic solids. Kepler believed that by fitting these five polyhedra between the six spheres carrying the planets, he could simultaneously explain the size of each planetary orbit and why there were just six planets.

Answer to Puzzle 1 The angles of the polygons that meet at each vertex of a tessellation must sum to 360°, or a complete rotation. At least three polygons must meet at a vertex. The only way to do this with regular polygons is with six equilateral triangles, as $6 \times 60° = 360°$; four squares, as $4 \times 90° = 360°$, or three hexagons, as $3 \times 120° = 360°$ (three pentagons would give: $3 \times 108° = 324°$, for instance).

Puzzle 2 A collection of polygons joined at their edges to enclose a volume of space without leaving any gaps is known as a polyhedron. The polyhedron is regular if the polygons are regular and all of the same type, with the same number meeting at each vertex. Why are there only five regular polyhedra, as shown in Figure 2.6?

HINT: In the square tessellation, four squares meet at each vertex, whereas in the cube only three squares meet at each vertex. In the equilateral triangle tessellation, six triangles meet at each vertex, whereas in a tetrahedron only three triangles meet at each vertex.

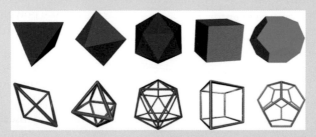

Figure 2.6 The five regular polyhedra. In each case the solid polyhedron is shown above its wireframe equivalent. From left to right: tetrahedron; octahedron; icosahedron; cube; dodecahedron.

After some experimentation, Kepler settled on the following scheme: between the orbits of Saturn and Jupiter was a cube, between Jupiter and Mars was a tetrahedron, between Mars and the Earth was a dodecahedron, between Earth and Venus was an icosahedron, and between Venus and Mercury was an octahedron (Figure 2.7). With a bit of flexibility with regards to its precise construction, Kepler was able to show that it fitted the dimensions of the solar system reasonably well, so well that he was convinced that he had discovered one of the most fundamental secrets of the universe. He would reveal this great discovery to the world in his first book *Mysterium Cosmographicum* (*The Secret of the Universe*), published in 1596. The twenty-four-year-old Kepler would send copies of the book to many of the leading astronomers in Europe, including Galileo and Tycho Brahe.

This book, Kepler's earliest publication, is infused with a strange mystical geometry and an enthusiasm to honour God by finding the blueprint of the universe. It is all quite mad from a modern perspective, but Kepler was now convinced that he must dedicate his life to astronomy and the quest for the true laws of nature.[4] Kepler's later writings would also be enveloped in strange mystical ideas and heavenly harmonies, but somehow amidst the wild fantasies would be some of the most profound ideas ever dreamt up. This was the start of an obsessive yearning to understand the structure of the universe that would ultimately lead to the dawn of the modern scientific era.

Answer to Puzzle 2 The angles of the polygons that meet at each vertex of a polyhedron must sum to less than 360°, and at least three polygons must meet at a vertex. The only way to do this with regular polygons is with three, four, or five equilateral triangles meeting at each vertex (which corresponds to the tetrahedron, octahedron, and icosahedron, respectively), three squares meeting at each vertex (which corresponds to the cube), or three pentagons meeting at each vertex (which corresponds to the dodecahedron).

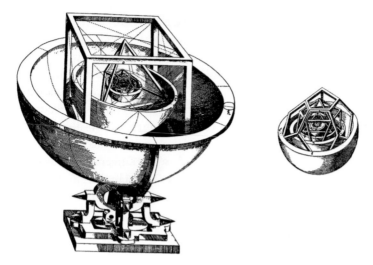

Figure 2.7   Kepler's polyhedral model of the solar system.

Kepler tried to persuade Frederick, Duke of Württemberg, to commission a cup designed in accordance with his model of the universe. In the words of Kepler's biographer Arthur Koestler:

> *Kepler went on to suggest that the various parts of the cup should be made by different silversmiths, and then fitted together, to make sure that the cosmic secret would not leak out. The signs of the planets could be cut in precious stones—Saturn in diamond, Jupiter in jacinth, the Moon a pearl, and so on. The cup would serve seven different kinds of beverage, conducted by concealed pipes from each planetary sphere to seven taps on its rim. 'The Sun will provide a delicious aqua vita, Mercury brandy, Venus mead, the Moon water, Mars a strong Vermouth, Jupiter a delicious new white wine', and Saturn 'a bad old wine or beer, whereby those ignorant in astronomical matters could be exposed to shame and ridicule'.*[5]

The Duke suggested that Kepler should first construct a model in copper. Kepler was strapped for cash, there was no way he could afford copper, so he spent a week manufacturing his design in coloured paper. Then he sent it off to the Duke, apologizing for its size. Ultimately, Kepler's plans for the grand silver goblet would come to nothing. But the spiritual quest to understand the mind of God and discover the architecture of the universe had only just begun. What Kepler craved were the reasons for the number, size, and motions of the planets. The task he set himself was to discover the celestial harmonies that he felt must govern the planetary motions. The only way in which he could fruitfully spend his time was in gathering proof of the heliocentric view of the universe. And there was just one man in Europe who had observed the planets with sufficient accuracy to settle Kepler's questions. This was the eccentric Danish nobleman, Tycho Brahe.

A Giant of Astronomy

Tycho Brahe (1546–1601) is one of the most incredible characters in the history of science. His life was dedicated to making the most precise measurements of the planetary positions that had ever been taken and he designed instruments of gargantuan proportions to achieve this unprecedented accuracy.

Tycho was born into one of the leading aristocratic families in sixteenth-century Denmark which, at the time, was a major northern European power holding what is now the southern tip of Sweden and thereby controlling the entrance to the Baltic Sea between Helsingborg and Elsinor Castle. The Danish Crown could exact duties on all the merchant vessels that pursued their lucrative trade through these waters. Tycho's father, Otto Brahe, and other close relatives were members of the Danish Council of State that formed the tier of government beneath the Danish king. Tycho was the first-born child and when a second child arrived about a year later, Tycho was kidnapped by his uncle Jorgen. Jorgen was a great sea captain who had distinguished himself in Denmark's naval encounters. He later became the vice-admiral of the Danish navy. Jorgen appears to have had an understanding with Tycho's father that he would adopt Tycho as soon as a second child was born. But when Tycho was not duly handed over, he was forcibly abducted by Jorgen. Tycho's parents eventually reconciled themselves to their child being raised by Jorgen and his wife, and Tycho seems to have benefited from the attention of two sets of parents.

In 1565, Jorgen was accompanying King Frederick II over a bridge outside Copenhagen after a heavy drinking session. The king was thrown from his horse into the water. Jorgen jumped in and rescued the king, but later died of pneumonia. The king remained grateful for Jorgen's sacrifice, and this might explain the enormous generosity that he later showed to Tycho.

As a young student, Tycho saw a partial eclipse of the Sun in Copenhagen on 21 August 1560. He was so impressed that it was possible to predict such events that it triggered his passion for astronomy. Tycho bought an astrolabe and began to make his own observations. A few years later, in 1563, still aged just sixteen, he observed the conjunction of Jupiter and Saturn and compared his own observations to the tables in the almanacs. All the published almanacs were woefully inaccurate in their predictions for the date of closest approach of the two planets. Indeed, the Alfonsine Tables, which dated back to the thirteenth century and were based on the Ptolemaic system, were out by a whole month, and even the Prutenic Tables, published just over a decade earlier in 1551 and based on the Copernican system, were out by several days. It was clear to Tycho that astronomy was in need of an overhaul, and that accurate and systematic measurements taken over a long period of time would be necessary to achieve this. With Jorgen's death, in 1565, Tycho inherited a substantial fortune, providing him with the funds to undertake this formidable task.

Duelling was endemic amongst the Danish aristocracy, as was heavy drinking. During the drunken Christmas celebrations of the following year, Tycho fell into a dispute with a distant cousin, Manderup Parsberg, and they retired outside to duel with rapiers. Tycho was slashed across the face and his nose was severed. For the rest of his life, Tycho would wear a prosthetic copper nose and, on special occasions, a nose forged from an alloy of silver

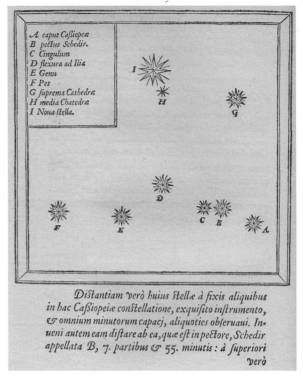

Figure 2.8 Illustration from *Stella Nova*. Stars F, E, D, B, G form the familiar 'W'-shaped outline of the constellation Cassiopeia. Star I is the new star.

and gold.[6] Remarkably, following this disfiguring and almost fatal clash, Tycho and Parsberg became lifelong friends.

Shattering the Crystal Spheres

In 1572, Tycho was amazed to see a new star appear in the constellation of Cassiopeia (Figure 2.8).

> On the 11th day of November in the evening after sunset, I was contemplating the stars in a clear sky. I noticed that a new and unusual star, surpassing the other stars in brilliancy, was shining almost directly above my head; and since I had, from

boyhood, known all the stars of the heavens perfectly, it was quite evident to me that there had never been any star in that place of the sky, even the smallest, to say nothing of a star so conspicuous and bright as this. I was so astonished at this sight that I was not ashamed to doubt the trustworthiness of my own eyes. But when I observed that others, on having the place pointed out to them, could see that there really was a star there, I had no further doubts—a miracle indeed, one that has never previously been seen before our time, in any age since the beginning of the world.

When it first appeared, Tycho estimated the new star to be as luminous as the planet Jupiter, but soon it brightened significantly until it even outshone Venus, the brightest of the planets. For about two weeks it could be seen in daylight. It then gradually faded and as it did so, it changed in colour from white to yellow to orange to pale red. After about sixteen months, it became too faint to see. Tycho measured the position of the new star against the background stars and attempted to determine its parallax by recording its position during the course of the year, but found no shift. This proved that the star must be more distant than any of the planets. Tycho published his observations in 1573 in *De Stella Nova* (*The New Star*).

As Tycho was keen to point out, his observations demolished Aristotle's claims that although the sub-lunar sphere was subject to decay, beyond the sphere of the Moon the heavens were eternal and unchanging. We now know that the *new star* was the result of a supernova explosion, in which a star blew itself apart in its almighty terminal blast.[7] Tycho's booklet *Stella Nova* was circulated throughout European learned circles and his work became famous.

A few years later, in 1577, Tycho dealt Aristotle's cosmos another crushing blow. In that year there was a bright comet with a blue-white head and a reddish tail. Tycho tracked its course from 17 November until January and demonstrated conclusively that the comet must be further away than the Moon.[8] Since Aristotle, the heavens were assumed to be composed of a nested sequence of crystal spheres whose function was to guide the planets in their

paths around the Sun. Tycho's analysis of the comet's motion
showed that the crystal spheres could not have any real physical existence. The comet had crossed the orbits of the planets and
would have shattered them.

King Frederick was duly impressed and rewarded Tycho on a
magnificent scale. This would be one of the most important in-
vestments in the history of science. His distinguished astronomer
was given the use of the island of Hven and grants of a number
of fiefdoms and benefices that would provide an income to build
and run an observatory there. Tycho would transform it into an
enchanted island rising out of the misty waters of the Oresund.

An Enchanted Isle

> Be not afeard; the isle is full of noises,
> Sounds, and sweet airs, that give delight and hurt not.
> Sometimes a thousand twangling instruments
> Will hum about mine ears; and sometime voices
> That, if I then had waked after long sleep,
> Will make me sleep again; and then in dreaming,
> The clouds methought would open, and show riches
> Ready to drop upon me, that when I waked
> I cried to dream again.
>
> WILLIAM SHAKESPEARE, *The Tempest*, ACT III, SCENE II

Hven is a small island, around 5 kilometres long by 2 kilometres
wide, within sight of Helsingborg in the straits between present-
day Denmark and Sweden. On its highest point, Tycho built his
observatory home, Uraniborg (Urania's Castle, Urania being the
muse of astronomy). Uraniborg was surrounded by a geometri-
cally laid out botanical garden aligned with the cardinal directions, where medicinal herbs were grown. At the centre of the
garden, Uraniborg was designed in an Italianate style reminiscent
of the Doge's Palace in Venice (Figure 2.9). There was an elaborate
clock tower. Two further towers were designed to hold Tycho's
huge instruments, and each was fitted with a conical wooden

Figure 2.9 The main building of Tycho's Uraniborg on the Island of Hven.

roof whose triangular sections could be removed to reveal the sparkling jewels of the night sky.

Figure 2.10 shows Tycho's giant quadrant and a mural that offers a glimpse into the rooms within Uraniborg. In the cellars were Tycho's alchemical laboratories, filled with furnaces and all the paraphernalia of the hermetic art. Above the laboratory was a circular library whose walls were lined with books, and in the centre of which was a great brass globe that was the focus of Tycho's painstaking research. During the course of his observations, Tycho systematically measured the positions of 777 of the most prominent stars in the sky, and steadily plotted each on this globe. He even built a paper mill and a printing press to publish his results. It has been estimated that Tycho's lavish expenditure accounted for as much as 1% of the total income of the Danish

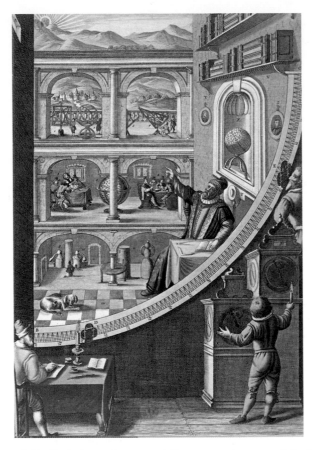

Figure 2.10 Tycho's quadrant partially surrounds a mural depicting Tycho seated in front of a cross-section of the Uraniborg observatory. In the basement of the observatory is Tycho's alchemical laboratory, above which is a library containing the giant globe and on the upper floor are Tycho's astronomical instruments. In the mural, Tycho points towards a narrow opening in the wall. In the foreground, Tycho is using the quadrant to measure the altitude of an object viewed through the opening. An assistant reads the time from a collection of clocks, while a second seated assistant records the observation.

crown.[9] Tycho is believed to have devised a communication system by which he could ring a bell in any room in the observatory to summon his assistants, almost as though he were summoning spirits to do his bidding.

Tycho's magical island attracted noble visitors from throughout Europe. In 1589, James VI of Scotland married Anne of Denmark, daughter of the Danish king, Frederick II, who had died the previous year. After visiting Denmark, James and his new wife were caught in a storm and took shelter on Tycho's island on 20 January 1590.[10] In 1601, James succeeded to the throne of England. It has been suggested that Shakespeare's play *The Tempest*, performed for James in 1611, recalls the visit of the monarch to Tycho's wondrous domain.

Tycho built a second observatory on the island and named it Stjerneborg or Star Castle. This observatory was constructed underground to improve the accuracy of the measurements, as it would be out of the wind and built on secure foundations that would not be subject to even the slightest disturbance. Tycho examined the origin of all possible inaccuracies in his observations and went to great lengths to eliminate them where possible. He was the first to systematically take account of these various sources of error and estimate their size. For instance, he measured the effects of refraction due to the atmosphere on the position of stars near the horizon so that he could correct for this displacement. He also took account of the distortions in the shape of his huge instruments as they bent slightly under gravity. This type of analysis is standard in modern science but it all began in Tycho's feudal barony on the island of Hven.

After accounting for all possible errors, Tycho was confident that his observations were accurate to between one and two minutes of arc.[11] (To put this in context, the diameter of the Moon is about thirty minutes of arc.) This was about ten times the accuracy achieved by any previous astronomer. For several decades Tycho used his enormous instruments to determine the positions of the stars and the planets with unrivalled precision.

Tycho believed that the true system of the universe was a halfway house between the systems of Ptolemy and Copernicus. In the Tychonic system (Figure 2.11), the Earth was stationary at the centre of the universe, orbited by the Moon, but all the planets orbited the Sun, and the Sun in turn orbited the Earth. This enabled Tycho to account for the fact that he was unable to observe any movement of the stars due to parallax. But the resulting system was an inelegant compromise.

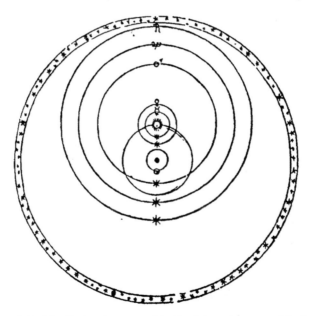

Figure 2.11 The Tychonic system. The Earth is at the centre. The Moon orbits the Earth. The Sun also orbits the Earth and all the planets orbit the Sun. Note that the fixed stars form a thin shell that encloses the universe. (By returning the Earth to a stationary position at the centre of the universe, Tycho had removed the need to assume that the stars must lie at an enormous distance in an essentially infinite universe.)

Following the death of King Frederick, Christian IV succeeded to the Danish throne and Tycho's world was turned upside down. The new king was hostile to Tycho and, as the years passed, relations between them deteriorated dramatically. Finally, in 1597, Tycho left his homeland, taking his entourage with him. After Tycho's departure, King Christian ordered his observatories to be demolished (Figures 2.12 and 2.13).

Tycho's immense reputation secured him the position of Imperial Mathematician to the Holy Roman Emperor in Prague. His task would be to prove his Tychonic system of the heavens and to compile a new up-to-date set of astronomical tables, superior to all previous tables, and dedicated to his new patron Rudolf II. Tycho was soon joined by the ideal assistant in Johannes Kepler. Tycho had been impressed by the originality and the mathematical ability shown by Kepler. From Kepler's

Figure 2.12 The grounds of Uraniborg.

Figure 2.13 The remains of Uraniborg on Hven.

point of view, the situation was perfect; he would be working for the only man in Europe whose observational data was accurate enough to answer his urgent questions about the structure of the universe.

The Battle with Mars

Kepler began work with Tycho in February 1600. Tycho was now fifty-three years old and Kepler was still just twenty-eight. Longomontanus, who was Tycho's chief assistant, had been assigned the task of determining the orbit of Mars, but he had made little progress. Kepler was so keen to demonstrate his abilities that he boasted he could complete the job within eight days, and so the task fell to him. Tycho kept a tight grip on his valuable observations and doled them out to his assistants on a need-to-know basis. This created much antagonism with his ambitious new mathematician. But at least Kepler now had some access to reliable and accurate measurements of the planetary positions. As Kepler began the painstaking analysis of the observations that Tycho had compiled, the scale of the challenge became clear.

Then suddenly everything changed. The larger-than-life figure of Tycho was struck down in a most surprising way. At a banquet held by Baron Rosenberg in Prague, Tycho consumed large quantities of food and drink but, according to Kepler's account, for the sake of etiquette he would not leave the banqueting table to go to the urinal. In Kepler's words, 'he put politeness before his health', with the consequence that he strained his urinary system. By the time that he had returned home, he found that he could not urinate at all. Kepler's account seems all the more remarkable as the mighty Tycho was very familiar with courtly habits and especially with eating and drinking on a grand scale. The feverish Tycho took to his bed, sleeping fitfully, if at all, and lapsing into delirium. He remained unable to urinate without experiencing excruciating pain. Eleven days after the banquet, he died. On his deathbed, Tycho pointed towards Kepler and then towards the heavens mumbling, 'let me not seem to have lived in vain'.

In 1901, exactly 300 years after Tycho's death, his tomb in Prague (Figure 2.14) was opened and his bones were examined. The skull clearly showed the wound from the duelling sword and close inspection revealed a greenish tinge produced by traces of Tycho's copper nose. Tycho's casket still contained strands of his ruddy beard, some of which were removed as souvenirs of the occasion. In the 1990s, samples of the beard hair were analysed, and the results revealed a darker side to the story of Tycho's strange death. The hair sample contained high concentrations of the toxic metal mercury, indicating that Tycho had consumed a large quantity of mercury in two doses, one about ten days before his death, which would coincide with the time of the banquet when he initially fell ill, and a second dose just a few hours before death. Furthermore,

Figure 2.14 Tycho's tomb in the Church of Our Lady Before Týn in Prague.

Tycho's symptoms are consistent with the effects of kidney failure due to mercury poisoning. So, who was responsible?[12]

Something Rotten in the State of Denmark?

Hamlet: Murder!
Ghost: Murder most foul, as in the best it is;
But this most foul, strange and unnatural.
WILLIAM SHAKESPEARE, *Hamlet*, ACT I, SCENE 5

The finger of suspicion has been pointed towards Kepler.[13] It is certainly true that he was the main beneficiary of Tycho's demise. Within days, an emissary arrived from the emperor to inform Kepler that he had been appointed the new Imperial Mathematician.

But there is a more credible suspect. A few days before Tycho was taken ill, a dissolute and impoverished distant relative, Erik Brahe, from the Swedish branch of the family, arrived in Prague. The historian Peter Andersen claims that the recently discovered diary of Erik Brahe implicates him in Tycho's downfall. According to Andersen, Erik was an agent sent by King Christian IV of Denmark to take Tycho's life. Furthermore, as a guest at Rosenberg's banquet, he had the opportunity to add a mercury compound to Tycho's goblet and another chance to administer the fatal dose on the evening before his death.[14] Anderson has speculated that the reason for King Christian's hatred of Tycho was that rumours were circulating in Denmark about an old love affair between the king's mother and Tycho which were casting doubts on Christian's legitimacy and therefore his right to the Danish throne. In November 2010, Tycho's body was exhumed again and underwent further analysis. The results were inconclusive as to the cause of Tycho's death and led to suggestions that the mercury in Tycho's body may have been due to his alchemical experiments rather than foul play. We may never know the true cause of Tycho's sudden death.

Wrangling over the inheritance began at once, with Tycho's heirs and entourage jockeying for position. Kepler realized that he must secure Tycho's four decades of planetary observations or his celestial journeys might end before they had barely begun. Kepler took possession of the grand chest containing the valuable data and refused to hand them over to Tycho's relatives. This was extremely fortuitous for the future of human civilization; Tycho's observations were in the hands of the one man who would be able to make use of them and raise the phoenix of modern science from the ashes of medieval mysticism.

Complex Gyrations

It is easy to take for granted the reckoning of time. The complex gyrations of the Sun, the Moon, and the planets barely elicit a passing thought. If we want to know the position of Mars, Jupiter, or Saturn in the night sky, we can consult a computer program that will give us their precise location.[15] But this knowledge was hard won. Kepler's official duty would now be to determine the true system of the universe and to compile the tables that would serve future generations of astronomers. They would be named the Rudolfine Tables in honour of Kepler's patron the Emperor.

With unhindered access to Tycho's data Kepler could concentrate all his efforts on revealing the secrets of the heavens. Copernicus had adopted the epicycles and other machinery of Ptolemy's system to produce a model that was no less convoluted than the systems of the ancients. Kepler believed that the universe was constructed on simple geometrical principles, so he recoiled in horror from the details of the Copernican system. Although he was a great admirer of Copernicus, he felt that his illustrious predecessor never knew the treasure that was within his grasp. Kepler realized that placing the Sun at the centre made much of the machinery redundant. He now had the data to construct a far superior system.

Kepler had a vision of a completely different universe—one in which the planets were controlled by natural forces. In his early writing, he refers to the planets being guided by spirits or minds. Later he was strongly influenced by the work of the English physician William Gilbert (1544–1603), who published *De Magnete* in 1600.[16] Gilbert described magnetism in terms of forces acting at a distance, and this was the view that Kepler adopted. For Copernicus, the role of the Sun was to provide light and heat from the centre of the universe. Kepler believed that the Sun was the heart of the mechanical system of the universe, and that its influence emanated outwards and controlled the motion of the planets. No-one had viewed the solar system in this way before. Kepler would never fully realize this vision, but it would guide him towards a clearer picture of the planetary motions than had ever been achieved before, paving the way for the modern understanding of the universe.

Kepler's task was monumental. It would consume six years of his life and would represent the pinnacle of his creativity. During the course of these investigations, he would systematically dismantle all the ancient machinery of epicycles, deferents, and equants that had been used during the two millennia between Hipparchus and Copernicus and reconstruct the planetary system using a totally new and elegant machinery of his own devising. At the start of the project astronomy was an arcane medieval art. With the publication of Kepler's results astronomy would become the first recognizably modern science. Kepler's achievement must be ranked as one of the greatest in the history of human thought. We are extremely fortunate that he left us an unembellished account of each step along the road to the correct solution, including all the dead ends and wrong turns.

Kepler had access to the most accurate and complete set of astronomical observations in existence. The observations recorded the position in the sky of each planet over the course of several decades. But disentangling the meaning of all this information in order to construct the geometry of the solar system would be far from easy. Just converting Tycho's raw data into a usable form was a major undertaking. The mathematical techniques that are taught in schools today were not available to Kepler. Algebra was in its infancy, coordinate geometry was unknown, and calculus did not exist. Consequently, the calculations that Kepler had to undertake were extremely lengthy and tedious. The only mathematical tools available to him were basic arithmetic, a little trigonometry, and the classical geometry contained in the works of the great ancient Greek mathematicians Euclid, Archimedes, and Apollonius. Beyond this, Kepler would have to devise his own methods.

Of Darkness and Light

We mark out the days in periods of darkness and light so it is natural that we should measure the length of a day by reference to the position of the Sun. The time between successive High Noons, when the Sun is due south (in the northern hemisphere), is not exactly twenty-four hours, as might be supposed; it varies gradually throughout the year. This is because the rate at which the Earth travels around the Sun varies. It moves fastest when closest to the Sun in early January, and slowest six months later. So the motion of the Sun across the sky is not exactly constant,[17] as was well known even in antiquity. By contrast, the Earth's rotation is extremely regular. It is therefore convenient to divide up the year into days of equal length that correspond to the period of the Earth's rotation.[18] This reasoning led astronomers to the notion of the mean Sun, which is defined by the position that the Sun would occupy if it crossed the sky at the same rate throughout the year, and this is why Greenwich Mean Time is so called.

When Copernicus constructed his heliocentric system, rather surprisingly, it was the mean Sun that he placed at the centre of the system and not the true Sun. This seems not to have mattered much to Copernicus as his aim was to produce a geometrical system that *saved the appearances*. But Kepler found this idea absurd, and it reflects his profoundly different view of the cosmos. Kepler felt that the Sun must somehow control the motion of the planets. This was a physical and causal relationship. It therefore made absolutely no sense to relate the motion of the planets to a fictitious entity such as the mean Sun. The first act of his revolution was to depose the mean Sun and set the true Sun at the heart of his system as the ruler of the planets.

There was an immediate pay-off to the promotion of the true Sun to its rightful place. The planetary orbits are almost in the same plane, like a collection of concentric circles, which is why the planets are confined within the band of zodiacal constellations. However, the alignment is not perfect; each orbit is inclined at a small angle to the others. For example, Kepler deduced from Tycho's data that the orbit of Mars is inclined by just under 2° relative to the orbit of Earth.[19] In the rickety Copernican system the orbit of each planet was completely independent. But Kepler could now see that the plane of each orbit intersected at the Sun (Figure 2.15), which reinforced his belief that the Sun was ruling the system of planets. This result is sometimes referred to as Kepler's Zeroth Law of Planetary Motion.[20] It was a first step towards a new system of the planets.

Eight Minutes that Shook the World

Kepler was keen to stress that Ptolemy and the other ancients who claimed to be preserving uniform circular motion in their theories had really been doing nothing of the sort. Kepler drew a diagram of the true path of Mars in the Ptolemaic system. He likened the looping epicyclic course of Mars to a pretzel and illustrated this with the diagram shown in Figure 2.16. As Kepler pointed out, the path was not circular, and the motion was not uniform. These ancient prejudices had forced astronomers into intellectual contortions, just like their planets.

Kepler dismissed the idea of epicycles as intellectually unsound. He believed that the correct solution must be much more elegant.

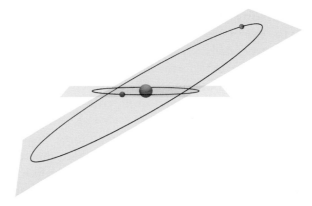

Figure 2.15 Kepler used Tycho's data to show that the plane of a planet's orbit is fixed in space and the Sun lies in the plane of each planet's orbit. This is sometimes known as Kepler's Zeroth Law (in contrast to the diagram, the orbits of the planets in the solar system are almost concentric).

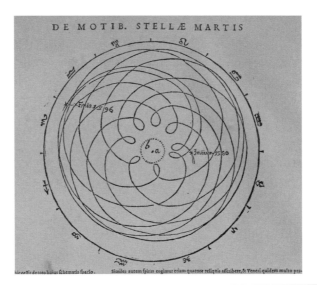

Figure 2.16 Mars's 'pretzel' pathways, illustrating the pretzel logic of Kepler's predecessors.

He set off in pursuit of Mars by assuming that its orbit was a perfect circle around the Sun. This would not work if the Sun were located at the centre of the circle, so he shifted the centre relative to the Sun. Initially, Kepler retained Ptolemy's notion of the equant point, a point close to the centre of the circle from which the motion of Mars would look uniform.

To trace out the path taken by Mars, Kepler searched through Tycho's data for the exact times and positions of Mars at opposition. This is when Mars is diametrically opposite the Sun in the sky, so Mars rises as the Sun sets and sets as the Sun rises. At these times the Sun, Earth, and Mars are in a straight line, which greatly simplified the calculations (Figure 2.17). Such oppositions occur every two years and seven weeks or so, and there were records of

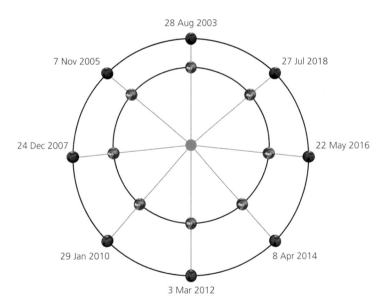

Figure 2.17 The figure shows the position of Mars when it reaches opposition on eight consecutive occasions. Kepler used Tycho's observations of the position of Mars at opposition to map out Mars's orbit.

ten oppositions in Tycho's data plus two that had been taken since Tycho had died. For instance, on 18 November 1580, at a time of 1 hour and 31 minutes, Mars was at opposition in Gemini and Tycho had recorded its exact position.

By taking four of these observations, Kepler could find a circular orbit and locate the equant point, thus producing a possible orbit for Mars. The next step would be to check whether this solution matched the other eight oppositions. If not, Kepler would have to make a slight adjustment to bring the orbit in line with these other observations. By trial and error, grinding through almost endless computations, Kepler gradually moved towards a solution that would fit Tycho's data. In his *New Astronomy*, he wrote: 'If this wearisome method has filled you with loathing, it should more properly fill you with compassion for me as I have gone through it at least seventy times at the expense of a great deal of time.'[21]

Finally, after these lengthy and tedious calculations, he obtained a circular orbit that fitted all twelve observations of the oppositions of Mars to within two minutes of arc. At this point all Kepler's predecessors would almost certainly have jumped for joy and considered that they had successfully completed their task— but not Kepler. He was determined to find the true description of the heavens. The orbit of Mars had to agree with all of Tycho's observations, not just those at opposition.

Kepler searched for any possible discrepancies and systematically calculated the course of Mars across the night sky according to his model. He then compared these predictions to Mars's actual position, as recorded in Tycho's observations, checking its position over the entire orbit, and not just at the oppositions which had been used in its derivation. Kepler's model was certainly close, but he could find observations that were as much as eight minutes of arc from the predicted position. This is the equivalent of about a quarter of the diameter of the Moon. No astronomer before Tycho had taken observations that were this accurate, but Kepler knew that Tycho had gone to great lengths to ensure that his observations were as accurate as they possibly could be. Kepler was certain that Tycho's observations were accurate to within two minutes of arc. He knew that if his model did not agree with Tycho then there could only be one conclusion—it was wrong—and he would have to start again. In the words of Kepler, 'these eight minutes showed the way to a renovation of the whole of astronomy'.[22]

> *After the divine goodness had given us in Tycho Brahe, so careful an observer, that from his observations the error of calculation amounting to eight minutes betrayed itself, it is seemly that we recognize and utilize in thankful manner this good deed of God's, that is we should take the pains to search out at last the true form of the heavenly motions.*[23]

Kepler was now convinced that he should discard all the rest of the ancient astronomical machinery and start again with a blank slate. Kepler's task would have been impossible without his deep conviction that the universe was constructed from simple geometrical principles.

Kepler Lays an Egg

In the Sun-centred model, the Earth orbits the Sun. This exacerbates the problem of finding the paths of the planets because it means that observations are taken from a moving platform. Kepler decided that he could only make progress with a better understanding of the Earth's motion around the Sun, so he now set out to determine the characteristics of the Earth's orbit accurately.

Kepler's idea for how this could be done was typically brilliant. He imagined viewing the Earth from Mars. He would choose a particular date in the Martian calendar and work out the position of the Earth on this date for successive Martian years. Mars would be at exactly the same position in its orbit for each observation, so any irregularities in the Martian orbit would be irrelevant. If the direction to Earth from Mars on that date was plotted for successive Martian years, this would map out the shape of the Earth's orbit.

It takes Mars 687 Earth days to orbit the Sun. By finding a sequence of measurements of the position of both Mars and the Sun made at intervals of exactly 687 days, Kepler could begin his task. On these dates, the Sun and Mars would be in the same positions. Only the Earth would have moved between the observations (Figure 2.18). By calculating the angle between the Sun

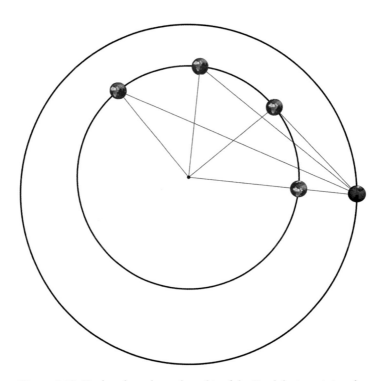

Figure 2.18 Kepler plotted out the orbit of the Earth by imagining that he was viewing it from Mars on the same Martian date in successive years. As shown in the figure, the position of Mars is the same for each observation, but the Earth has moved around its orbit.

and Mars from each observation, it would be possible to plot out the path of the moving Earth. The calculations would be just like working out the aspects for an astrological chart.

Kepler succeeded in finding the precise course of the Earth's orbit around the Sun. He could now concentrate on Mars.[24] Now, when he plotted out the path of the red planet, he thought he had the answer. It was shaped like an egg! It seemed as though the orbit was rounded when Mars was furthest from the Sun, and narrower when close to the Sun.

Heavenly Movement

Kepler struggled on with his egg-shaped orbit for several years without ever being convinced that he had solved the problem. Finally, it all fell into place. Kepler realized that he had been holding the answer in his hands all the time. One of the most famous books of mathematics to survive from antiquity was the *Conics* by Apollonius of Perga. Kepler was very familiar with the book and had used it in his work on optics.[25] Apollonius examined the properties of the geometrical figures that are produced by slicing through a cone (Figure 2.19). For instance, if a cone is cut by a horizontal slice, the cross-section is a circle. If the cone is sliced at an angle, however, the shape of the cross-section is an ellipse. Increasing the angle of the slice increases the eccentricity of the ellipse. If the slice is parallel to the edge of the cone, then the figure will be open-ended. It is known as a parabola.

There are other ways to construct these figures, as Apollonius showed. It is easy to define a circle. The circle consists of all the point whose distances from the centre are equal to the radius of the circle. An ellipse can be defined in a similar way but, this time, two points are required. Each is known as a focus of the ellipse. Take a string and attach one end to each focus. The ellipse consists of all the points that can just be reached when this string is stretched taut, as illustrated in Figure 2.20. If the two foci are moved further apart, the eccentricity of the ellipse increases. If

Figure 2.19 Conic sections. From left to right: circle, ellipse, parabola, hyperbola.

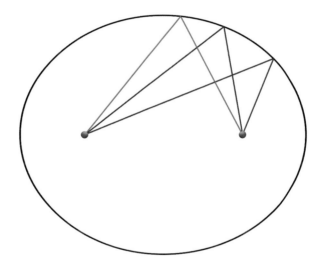

Figure 2.20 An ellipse can be constructed by specifying two interior points—each known as a focus of the ellipse—then, with a piece of string fastened to each focus, every point that can just be reached by making the string taut is on the ellipse. In other words, take any point on an ellipse and draw straight lines to the two foci; the sum of the lengths of these two lines is the same for all points on the ellipse. Three examples of such pairs of lines are shown in the illustration.

they are moved together, the eccentricity of the ellipse decreases until it becomes a circle when the two foci merge together as the same point—the centre of the circle.

One of the interesting properties of an ellipse is that if there were a lightbulb at one focus, then all the light that it emits would reflect off the ellipse and converge at the other focus. This is why Kepler originally used the name *focus* for these points. Furthermore, if one focus is removed all the way to infinity, then the ellipse becomes a parabola. To a very good approximation any light entering the parabola from a distant light source will appear to arrive from the focus at infinity. It will therefore reflect off the parabola and converge at the other focus. This is the basis for the design of the reflecting telescope. The *parabolic* mirror in such telescopes is shaped like the end of a parabola rotated around its axis.

Kepler finally realized that the orbit of Mars must be an ellipse, with the Sun positioned at a focus of the ellipse. This was a remarkable result. It is known as Kepler's First Law of Planetary Motion. The ellipse might not be quite as symmetrical as a circle but it was a wonderful second best. It was certainly vastly better than an ill-defined egg shape. Much more importantly, it accurately fitted the facts.[26] Kepler went on to show that all the other planets also follow elliptical orbits around the Sun. As it turns out, it was quite fortuitous that Kepler had undertaken his epic struggle with Mars; none of the other planets could have provided the data for this discovery. Mercury's orbit is very eccentric, but it is so close to the Sun that it is only visible in the twilight, and this means that it is difficult to make accurate observations. The orbit of Venus has a very low eccentricity, so it would have been extremely difficult to distinguish it from a circle. The outer planets Jupiter and Saturn take so long to orbit the Sun that several lifetimes would be needed to collect sufficient data. Mars is the only planet that could have offered Kepler the information he needed to deduce his First Law.

Figure 2.21 Johannes Kepler, painted in 1610 by an unknown artist.

The New Astronomy

The shape of the planetary orbits went only part way to answering Kepler's questions. He also demanded to know the rules that determine the speed of each planet. Kepler (Figure 2.21) knew that a planet would move more quickly when it was close to the Sun. This seemed reasonable, because he believed that the Sun, as the centre of the heliocentric system, must somehow be propelling the planets around it. Tycho's observations appeared to confirm

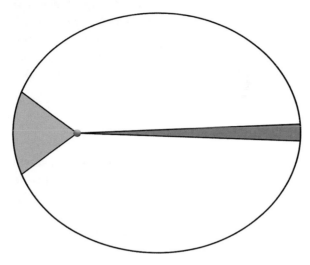

Figure 2.22 Kepler's Second Law states that a planet sweeps out sectors of its orbit of equal area in equal periods of time. The illustration shows two such sectors.

this idea but Kepler believed that it should be mathematical and quantifiable. What he wanted was a precise geometrical or arithmetical expression for the way in which a planet's speed changes as it approaches and recedes from the Sun, and he was determined to find one. He found the answer even before he knew the correct shape of the orbits.

After rejecting the equant point, Kepler discovered what he thought was a useful, but approximate, way to determine the motion of a planet.[27] He later realized that his rule held exactly. It goes as follows: a planet sweeps out sectors of its orbit of equal area in equal periods of time (Figure 2.22). Kepler's laws are numbered in logical order and not in the order in which he found them, so this law is known as Kepler's Second Law of Planetary Motion.

Kepler's revolutionary system of the universe was revealed to the world in 1609 in his book *Astronomia Nova* (*The New Astronomy*).

This was Kepler's masterpiece, the product of many years of intensive work. It was the first book ever written that would explain natural processes by comparing accurate measurements to a mathematical and deterministic theory. It has become the model for how theoretical physics works and its importance for the future of science is unparalleled.

The Harmony of the World

John Napier (1550–1617) was a Scottish aristocrat who inherited the title 8th Laird of Merchiston (Figure 2.23). He was born in the Merchiston Tower in Edinburgh, which is now part of Edinburgh Napier University.

Napier was a mathematician who dedicated himself to developing new and better methods of calculation to ease the burden on navigators and astronomers. In 1614, Napier published *Mirifici Logarithmorum Canonis Descriptio*, the first book of logarithms ever published. Part of the book was an explanation of the theory behind logarithms, and the rest consisted of ninety pages of tables. The key feature of logarithms was that they enabled mathematicians to simplify their calculations dramatically by converting multiplication and division sums into addition and subtraction sums. To multiply two numbers together, an astronomer would look up each number in the tables of Napier's book and read off the value of the logarithm in each case. These two logarithms would then be added together. The astronomer would then look in another table to find the number whose logarithm was equal to this result, and this number would be the answer to the multiplication sum.

For an astronomer such as Kepler, this offered an enormous saving in labour. Kepler obtained a copy of Napier's book within two years of its publication. He was so grateful for Napier's method that he dedicated his *Ephemeris* to the Scot. Throughout his life, Kepler searched for the harmonic laws of the heavens—rules that would explain the motion of the planets inspired by the

Figure 2.23 John Napier, 8th Laird of Merchiston.

Pythagorean notion of the music of the spheres. In 1618, he had a sudden revelation in which he saw a relationship between each planet's orbital period and its distance from the Sun. Napier's logarithms almost certainly played a significant role in this discovery. What Kepler discovered was his harmonic law, which states that the square of the orbital period of a planet is proportional to the cube of its orbital radius.

The following example illustrates how the relationship works. Take the radius of the Earth's orbit to be one astronomical unit, then the radius of Saturn's orbit is about nine astronomical units. The period of the Earth's orbit is one year. Kepler's harmonic law says that if we cube the radius of Saturn's orbit in astronomical units and then take the square root, we will find the period of Saturn's orbit in years. Nine cubed is 729. The square root of 729 is 27. By this reckoning Saturn should orbit the Sun once every twenty-seven years. In fact, we have rounded the figures to make the calculations simpler for the purposes of this illustration. (The radius of Saturn's orbit is 9.537 astronomical units, and its period is 29.45 years. It is worth putting these more precise numbers into a calculator to see just how accurate the relationship is.)[28]

Stumbling across such a relationship amidst a morass of figures would have been quite unlikely. With the assistance of logarithms, however, the task becomes much simpler. After taking logarithms of the orbital periods and orbital radii of the planets, the relationship would be that twice the logarithm of the orbital period is proportional to three times the logarithm of the orbital radius. If these figures are plotted on a graph, the data for each planet sit at a point on a straight line. Although Kepler did not draw such a graph, the numbers were so familiar from his endless calculations that the relationship must have leapt out at him. Kepler's harmonic law is also known as his Third Law of Planetary Motion.

Puzzle 3 A hypothetical asteroid orbits the Sun at four times the distance of the Earth. How long does it take the asteroid to orbit the Sun?

Answer to Puzzle 3 Kepler's Third Law is the key to answering this puzzle—the cube of the radius of the orbit is proportional to the

continued

Continued

---

square of the orbital period. The cube of the radius of the asteroid's orbit is sixty-four times the cube of the radius of Earth's orbit. To find the period of the orbit, we take the square root to give eight times the period of Earth's orbit, or eight years.

Measuring the Shadows of the Earth

Kepler published the long-awaited Rudolfine Tables in 1627, setting a new standard for the accuracy of astronomical tables. Far more important, though, was the conceptual revolution that Kepler initiated by showing that the planetary orbits are ellipses. Kepler's research can be succinctly distilled into his laws of planetary motion. These are:

0   The plane of a planet's orbit is fixed in space and the Sun is situated in this plane.
1   Each planet orbits the Sun in an ellipse, with the Sun located at a focus of the ellipse.
2   Each planet sweeps out sectors of its orbit with equal area in equal intervals of time.
3   The cube of the length of the major axis of a planet's elliptical orbit is proportional to the square of its orbital period.

Kepler believed that the universe is controlled by the operation of forces and, more specifically, that the Sun produces the driving force that maintains the planets in their orbits. His speculations about how this might work were unsuccessful. His best guess was that the rotation of the Sun produces an influence that sweeps the planets round. He also toyed with the idea that the force might be some kind of magnetic influence. This is, of course, completely wrong. It might seem obvious to us today that the force that holds the planets in orbit is another manifestation of the force that holds us to the ground, but this was far from obvious, even to a researcher with the imagination and tenacity of Kepler. Nonetheless, the laws of planetary motion that Kepler discovered would play a decisive role in enabling Sir Isaac Newton to discover his laws of motion and his universal law of gravitation, as we will soon see.

Kepler died on 15 November 1630 and was buried in the Bavarian city of Regensburg. The cemetery was destroyed by a Swedish army just two years later during the Thirty Years War, so Kepler's tombstone has been lost. But the epitaph that Kepler wrote for himself has survived.[29]

I measured the Heavens, now I measure the shadows of the Earth.
The mind belonged to Heaven, the body's shadow lies here.

3

The Magic Spyglass

He scarce had ceas't when the superiour Fiend
Was moving toward the shore; his ponderous shield
Ethereal temper, massy, large and round,
Behind him cast; the broad circumference
Hung on his shoulders like the moon, whose Orb
Through Optic Glass the Tuscan Artist views
At Ev'ning from the top of Fesole,
Or in Valdarno, to descry new Lands,
Rivers or Mountains in her spotty Globe.
JOHN MILTON, *Paradise Lost*, BOOK I, LINES 283–291, 1667.

The Starry Messenger

The year 1609 is not remembered as a particularly significant year; it is not engraved into our consciousness in the way that 1066, 1492, 1815, or 1945 might be. But if any year could be singled out as the beginning of the modern world, then this year probably has the best claim to that status. As we have seen, 1609 was the year in which Kepler published his *New Astronomy*, the book that overthrew 2000 years of astronomy and replaced it with a modern understanding of the planetary motions. This revolution in theoretical astronomy coincided with an equally momentous transformation of observational astronomy. In 1609, the Italian scientist Galileo Galilei (1564–1642) heard news that a remarkable optical instrument had been invented in the Netherlands. Galileo began to experiment with lenses and soon constructed his own instrument—a spyglass or telescope. With further experimentation, he was able to increase the magnification of the telescope, and he then began to use it to observe the night sky as never before. What he saw was astonishing.

Within a few months, in March the following year, Galileo published a short booklet describing his discoveries. It came like a bolt out of the blue. *The Starry Messenger*, as it was called, began with a triumphant fanfare:

> *THE STARRY MESSENGER revealing great, unusual, and remarkable spectacles, opening these to the consideration of every man, and especially of philosophers and astronomers; AS OBSERVED BY GALILEO GALILEI Gentleman of Florence Professor of Mathematics in the University of Padua, WITH THE AID OF A SPYGLASS lately invented by him, in the surface of the moon, in innumerable Fixed Stars, in Nebulae, and above all in FOUR PLANETS swiftly revolving about Jupiter at differing distances and periods, and known to no one before the Author recently perceived them and decided that they should be named THE MEDICEAN STARS.*[1]

Galileo made sensational discoveries everywhere he looked in the night sky. He found that, contrary to accepted wisdom and the claims of the followers of Aristotle, the Moon is not a perfectly spherical globe but, like the Earth, it is covered with mountains and crevasses. Throughout the heavens, he saw multitudes of stars that were too faint to see with the naked eye. The Milky Way was resolved into vast swarms of faint stars for the first time.

But Galileo's greatest discovery, as he presaged in his frontispiece, was the existence of four satellites of Jupiter. His name, the Medicean Stars, did not catch on, however, and these satellites are known today as the Galilean moons of Jupiter: Io, Europa, Ganymede, and Callisto. Galileo first saw three of the moons on 7 January 1610. On this first night, he assumed that they were background stars, although he noted that they were all in a line right on the ecliptic. After another couple of nights, it was clear that they were actually satellites dancing around the globe of Jupiter. Galileo drew the position of the satellites night by night

(Figure 3.1) over the course of the next two months running up to the publication of *The Starry Messenger*.

As Galileo pointed out, the new moons were another blow to the traditional view of the heavens; there was clearly much in the universe that was unknown to the ancients and moderns alike (Figure 3.2). It is very easy to see the Galilean moons of Jupiter with modern equipment. They are revealed by a casual view of Jupiter with even the smallest telescope or a pair of binoculars.

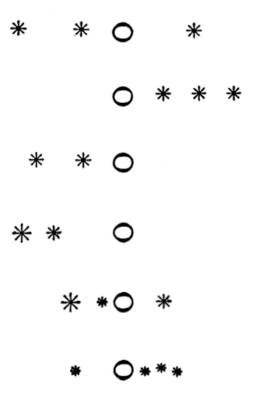

Figure 3.1 Galileo's drawings of the moons of Jupiter from *The Starry Messenger*. From top to bottom, as observed with his spyglass on 7, 8, 10, 11, 12, and 13 January 1610.

Figure 3.2 Galileo showing the Doge of Venice how to use the telescope. Fresco by Giuseppe Bertini (1858), Villa Andrea Ponti, Varese,
Italy.

Galileo was born in Pisa in 1564. He was the son of Vincenzo
Galilei who was a famous musician and music theorist with a
healthy scepticism about established authority. Galileo became
an accomplished lutenist, and his sense of rhythm and timing
may have played a part in some of his most famous discoveries. In
1581, Galileo enrolled at the University of Pisa to study medicine.
Galileo's biographer, Vincenzo Viviani (1622–1703), tells the story
that while Galileo was in the cathedral in Pisa, he noticed that

the great chandelier was swinging back and forth. Using his pulse to time the chandelier, he realized that the time taken to complete a full swing was the same, whatever the size of the swing (as Galileo later demonstrated in his experiments with pendulums, this is only strictly true for small oscillations).[2]

Galileo realized that his discovery might have a very important application. The swing of the pendulum could be used to measure time. Many years later Galileo designed a pendulum clock. His son Vincenzo was charged with its construction, but it was never completed. Galileo's idea would eventually come to fruition when the Dutch physicist Christiaan Huygens (1629–1695) constructed the first pendulum clock in 1656. This was a great technological breakthrough and pendulum clocks remained the most accurate timekeepers until the invention of the atomic clock in the 1930s.

Escaping the Dark Labyrinth

Philosophy is written in this immense book that stands ever open before our eyes (I speak of the universe), but it cannot be read unless we first learn the language and recognise the characters in which it is written. It is written in mathematical language, and the characters are triangles, circles and other geometrical figures, without the means of which it is humanly impossible to understand a word; without these philosophy is a confused wandering in a dark labyrinth.

GALILEO, *The Assayer* (1623)[3]

In 1602, while Kepler was looking up to the heavens and wrestling with the orbit of Mars, Galileo began a systematic investigation of motion closer to home. Galileo rejected the traditional explanations of the world dating back to Aristotle. He believed that the universe operated along mathematical principles that could be revealed through measurement and experiment. It sounds so obvious from our vantage point but prior to Galileo, the world was the subject of philosophical and theological arguments that were very difficult to pin down, made by scholars who rarely felt the need to check the assertions that they made.

Galileo is said to have dropped a wooden ball and a cannonball from the Leaning Tower of Pisa in order to demonstrate that they hit the ground together. Galileo may or may not have impressed his patrons with a theatrical demonstration such as this. What he definitely did do was construct ramps down which he could roll balls made of various materials. This slowed the balls down, making their motion much easier to analyse.

One of the most difficult problems that Galileo faced was how short periods of time could be accurately measured. Galileo solved the problem by designing a water clock from which a steady trickle of water would flow. He could then collect and weigh the water in order to quantify the amount of time that had elapsed. Using this simple equipment, Galileo showed that it did not matter what material the balls were composed of—they all rolled down his slopes in the same period of time.

Varying the angle of the slope led Galileo to the conclusion that the balls would only come to a halt due to the small amount of friction to which they were subject. He believed that if a ball were set in motion on an ideal horizontal track from which friction had been completely removed, the ball would continue onwards at the same speed indefinitely.[4] Newton would call this property of matter *inertia*, borrowing the term from Kepler, who believed that matter displays an inherent laziness. It would be key to Galileo's arguments for why we do not feel the rotation of the Earth or motion at a constant velocity, as he describes so eloquently in the following passage.

Shut yourself up with some friend in the main cabin below decks on some large ship, and have with you there some flies, butterflies, and other small flying animals. Have a large bowl of water with some fish in it; hang up a bottle that empties drop by drop into a wide vessel beneath it. With the ship standing still, observe carefully how the little animals fly with equal speed to all sides of the cabin. The fish swim indifferently in all directions; the drops fall into the vessel beneath; and, in throwing something to your friend, you need throw it no more strongly in one direction than another, the distances being equal; jumping with your feet together, you pass equal spaces in every direction. When you have observed all these things carefully (though doubtless when the ship is standing still, everything must happen in this way), have the ship proceed with any speed you like, so long as the motion is uniform and not fluctuating this way and that. You will discover not the least change in all the effects named, nor could you tell from any of them whether the ship was moving or standing still.[5]

What Are the Odds on That?

Puzzle 4 What is the sum of the first twenty odd numbers?

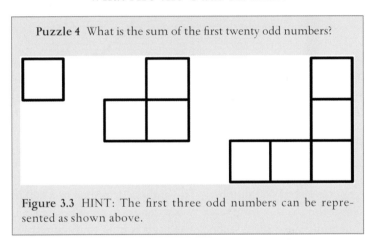

Figure 3.3 HINT: The first three odd numbers can be represented as shown above.

Galileo's next step was to examine the rate at which the balls ran down the slope. He found that the speed of a ball increases as it travels downwards, and it does so in a quite regular way. If the distance travelled by a ball in one second is taken as the unit of distance, then in the next second, the ball travels three units, in the third second it travels five units, in the next second seven units, and so on (Figure 3.4). This was a remarkably simple rule and the key to understanding motion close to the Earth. Galileo believed that he had found one of the arithmetical secrets of the cosmos. He called it his law of odd numbers.[6]

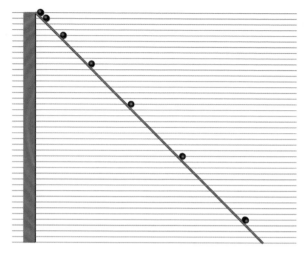

Figure 3.4 The diagram shows a ball rolling down a slope under the influence of gravity. The position of the ball is shown at equal intervals of time. The ball obeys Galileo's law of odd numbers. (The slope is much steeper than it would have been in Galileo's experiments.)

This puzzle illustrates a result that Galileo would later derive from his law of odd numbers. The distance travelled by a ball in one second is one unit; after two seconds, the total distance is four units; after three seconds the total is nine units; after four seconds the total is sixteen units and so on. In general, the total distance travelled by a ball grows as the square of the time that it is travelling down the slope.[7] In modern terms, what Galileo had discovered was that the rate at which the balls were moving was increasing steadily: in other words, they were undergoing a constant acceleration due to the force of gravity.

Answer to Puzzle 4 Odd numbers can be represented as 'L' shaped collections of squares, as shown in the hint. The 'L' shapes fit together to form a square thus:

continued

Continued

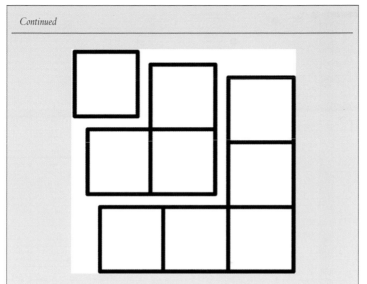

Figure 3.5 From the diagram it is clear that the first three odd numbers add up to $3^2 = 9$ and, in general, the sum of the first n odd numbers is equal to n^2. The sum of the first twenty odd numbers is therefore $20^2 = 400$.

The next step was to consider two motions at once. Galileo arranged for the balls to roll down slopes and then shoot off his table so he could analyse their trajectory through the air until they hit the floor. He found that the horizontal and vertical motions of the balls were completely independent. The path followed by a ball was a combination of these two separate motions. In the horizontal direction, the ball continues in a straight line at a constant speed, covering equal distances in equal periods of time, as Galileo had already deduced. Vertically, the motion would obey the odd numbers rule corresponding to a constant acceleration. As we have seen, this means that the distance covered in the vertical direction increases with the square of the time. The path

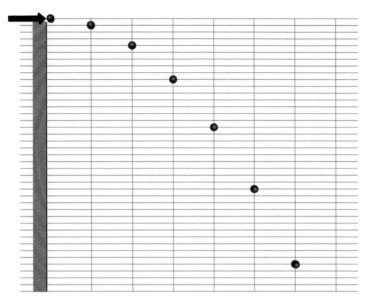

Figure 3.6 The diagram shows the trajectory of a ball that is pushed horizontally from the top of a platform. The path followed by the ball is a parabola.

followed by the ball is shown in Figure 3.6. The shape of the path is known as a parabola. Galileo had now found a geometrical law of nature.

A parabola can be regarded as an extremely eccentric ellipse. The greater the distance between the two foci of an ellipse, the greater its eccentricity; in the limiting case where one focus is taken all the way to infinity, the resulting figure is a parabola. There is a clear connection between Kepler's discovery that the planets move in elliptical orbits around the Sun and Galileo's discovery that projectiles travel in parabolic arcs,[8] but for many years no-one saw this connection. Galileo's results were published after Kepler's death and Galileo does not seem to have paid any attention to Kepler's books.

Galileo wrote in Italian, which was very unusual in the seventeenth century, when almost all learned discourse was carried on in Latin. He recorded the results of his experiments in motion and mechanics in his book *Two New Sciences*, published in 1638.[9] Galileo is famous for his eloquent and persuasive prose. His literary style certainly helped his ideas and discoveries to achieve a wide circulation.

Unfortunately, scientific enquiry in Galileo's homeland was brought to a juddering halt and the epicentre for original thought moved northwards from the Mediterranean to countries where scientific speculation could be pursued without hindrance. In 1633, Galileo was tried and found guilty by the Inquisition of arguing for the sun-centred cosmology of Copernicus. He was accused of heresy and required to *abjure, curse, and detest* the views that he had expressed. Galileo was sentenced to imprisonment, but the sentence was later commuted to house arrest in his villa in Arcetri just south of Florence, which he endured for the rest of his life. Within a few years he was blind due to a combination of cataracts and glaucoma, but he continued to work (Figure 3.7).

In 1638, the seventy-four-year-old Galileo was visited by a young poet who was travelling around Italy. His name was John Milton, and he would become the greatest English epic poet. In his masterpiece *Paradise Lost*, Milton makes several references to Galileo and his astronomical discoveries, including the extract that opens this chapter. Galileo is the only contemporary figure that Milton mentions in his great work. Galileo died in 1642, four years after his meeting with Milton.

Milton became an ardent champion of the freedom of the press. During the English Civil War, he would find some of his own writings condemned and censored. In 1644, Milton attacked such censorship in the *Areopagitica*, a passionate defence of the rights of an author to publish his work uninhibited by the threat of

Figure 3.7 Galileo and his last disciple Viviani by Tito-Giovanni Lessi (1892).

persecution. In support of his case, he made this reference to his visit to Italy:

> There it was that I found and visited the famous Galileo, grown old a prisoner to the Inquisition, for thinking in astronomy otherwise than the Franciscan and Dominican licensers thought.[10]

Milton's short treatise played an important role in the development of modern views on freedom of expression, and it had a significant influence on the constitution of the United States when it was drawn up in the following century.

Figure 3.8 The 2004 transit of Venus.

Venus in the Face of the Sun

On 8 June 2004 and 6 June 2012, the planet Venus passed across the face of the Sun[11] (Figure 3.8). These were only the sixth and seventh occasions on which this rare spectacle had ever been observed. The first person to witness such an event was a young astronomer called Jeremiah Horrocks, living in an obscure Lancashire hamlet, who correctly forecast that it would occur on 24 November 1639.[12] This was just a year after Milton's visit to Galileo, and three years before Galileo's death.

Planetary positions can now be predicted many years in advance, and this information is readily available on websites such as www.heavens-above.com. So, we know that the next transit of Venus will not occur until 11 December 2117.[13] Although Horrocks is not very famous today he was a brilliant mathematician and astronomer, and his observation of the transit of Venus was not simply a fluke. Indeed, he was probably the first person to realize that the path of a ball thrown through the air might be controlled by the same agency as the orbits of the planets around the Sun. The Newtonian revolution was partially built on his astronomical research, so his legacy has proved to be very valuable indeed.

But who was Jeremiah Horrocks?

Jeremiah Horrocks

Jeremiah Horrocks was born in 1618 into a family of watchmakers in Toxteth Park in Liverpool. Little is known about his earliest years, but he was no doubt captivated by his family's interest in the measurement of time and its connection to the whirling cycles of the heavens. In 1632, at the age of just fourteen, he was sent off to Emmanuel College in Cambridge, which was a leading centre for the study of mathematics and astronomy. At the college, he befriended John Wallis (1616–1703) who, although a couple of years older, entered the college in the same year. Wallis would become one the great mathematicians of the age, and one of the founders of the Royal Society. Emmanuel College gave Horrocks access to books on astronomy and contact with some of England's leading astronomers. It is probable that Horrocks learnt his observational skills from Samuel Foster, the author of treatises on quadrants and sundials. Foster had already been at Emmanuel for sixteen years when Horrocks arrived, and he would go on to become the Gresham Professor of Astronomy.

In 1635, Horrocks, still only seventeen years old, left Cambridge to return to his native Lancashire. It was just a quarter of a century since Galileo had rocked the learned world with *The Starry Messenger*. Galileo's spyglass had revealed numerous features of the universe that had never been seen before; the only way that someone of modest means could acquire a telescope to see these wonders at first hand was to make their own, and this is exactly what Horrocks did. Horrocks confirmed for himself all the discoveries of

Galileo. But he was no ordinary observer. What made him special was that he was also an excellent mathematician. Crucially, he had the ability to understand the various astronomical tables that were in existence and to perform the calculations that would predict the positions of the planets.

Initially, Horrocks used the recent astronomical tables compiled by the Dutch astronomer Philippe van Lansberge and published in 1632. These tables were based on Copernicus's model of the solar system, however, and they completely ignored the huge progress made by Kepler. Horrocks soon realized that Lansberge's tables were hopelessly inaccurate, despite the extravagant claims of their author. Horrocks obtained a copy of Kepler's Rudolphine Tables and immediately recognized their clear superiority.

Horrocks was a meticulous and accurate observer, but he was also an accomplished theorist who understood Kepler's ideas better than any of his contemporaries. Kepler had died in 1630, and Horrocks feared that his great advances might be lost, with astronomers such as Lansberge reverting to the crude and complicated epicyclic systems of the past. This made him all the more keen to demonstrate the advantages of Kepler's system. Horrocks confirmed through his own observations that Kepler's ideas worked to a very good approximation, and it was not long before he went even further.

Kepler's universe was built on a divine harmony in which the imperial Sun controlled its family of planets. According to Kepler, the Sun emanated a force that generates the motion of the planets, somehow sweeping them around like a broom. But ultimately Kepler failed in his attempts to find a mechanical explanation of how this solar force might work.

Horrocks recognized that the Sun was indeed controlling the movement of the planets, but he could also see that the motion of the planets could not be explained by the action of the Sun alone. His investigations suggested that there must be a force acting between all celestial bodies. And this was not merely philosophical speculation; it was based on a detailed analysis of the historical records of planetary positions, combined with his own observations.

For instance, when Horrocks analysed the paths of Jupiter and Saturn, the two largest planets in the solar system, it was clear that they affect each other's motion; there seemed to be some sort of mutual attraction. Jupiter orbits the Sun on the inside of Saturn, and every twenty years it overtakes the outer planet. Horrocks showed that as Jupiter closes in on Saturn it speeds up, but when it has passed Saturn it slows down. There was a similar effect on Saturn. As Jupiter was catching up Saturn appeared to slow down, then once Jupiter had passed by Saturn appeared to speed up again. It was as though the two planets were pulling on each other. This small effect was the key to an understanding of celestial mechanics that was quite different to Kepler's. Indeed, Horrocks was well on the way towards a much more sophisticated view of the force of gravity. Newton would later definitively demonstrate that all massive bodies exert a gravitational attraction on all other massive bodies, and this would transform our understanding of the universe.

Horrocks also applied Kepler's ideas to the Moon and was the first to show that the Moon follows an elliptical orbit around the Earth. But the Moon's motion proved to be rather difficult to model accurately. One of the effects that Horrocks pinned down is that the Moon's orbit precesses. This means that if a line is drawn joining the point at which the Moon is closest to the Earth to the point at which the Moon is furthest from the Earth, this line does not remain fixed in space, as Kepler's laws would suggest, but it gradually changes direction (Figure 3.9). In just under nine years,[14] the axis of the Moon's orbit completes a full circuit around the Earth (Figure 3.10).[15] Horrocks also discovered other periodic variations in the Moon's motion and determined their rate of change. Moreover, Horrocks was the first to realize that the

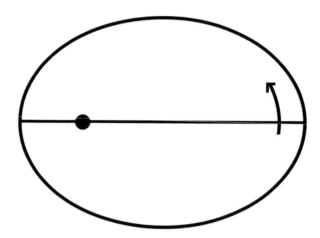

Figure 3.9 The Moon's orbit around the Earth is an ellipse to a very good approximation but, as Horrocks discovered, the direction of the axis of the ellipse gradually changes. This is known as precession. A full circuit is completed in about nine years. (The eccentricity of the orbit is greatly exaggerated in the illustration.).

reason why the Moon's course is so difficult to predict is that it is simultaneously affected by an attraction to both the Earth and the Sun. This was an idea that was half a century ahead of its time.

Horrocks also believed that there might be a connection between the gravitational force that we feel on Earth and the motion of the planets. This was potentially a huge advance on Kepler and would become a key ingredient in the Newtonian synthesis. Horrocks even suggested that a stone thrown into the air would follow an elliptical path, just like the Moon or a planet, if it were not brought to a premature halt by hitting the ground. In search of further insight into this problem Horrocks experimented with pendulums.

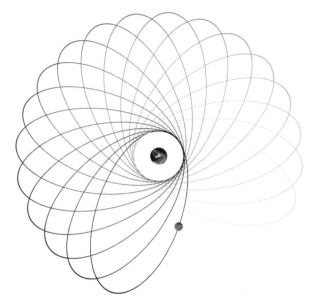

Figure 3.10 Diagram showing the precession of the axis of the Moon's orbit around the Earth. (The eccentricity of the orbit and the rate of precession are exaggerated.).

The Pendulum

England swings like a pendulum do.

ROGER MILLER (1965)

Kepler had discovered the shape of the planetary orbits and how the speed of a planet changes as it moves around its orbit—speeding up as it approaches the Sun and slowing down as it recedes from the Sun. He had demolished the astronomical systems of the ancients and replaced them with a much simpler and more accurate system. This was one of the great achievements in the history of science but, as Kepler well knew, it was only a halfway step to a new physics. Kepler's system explained how the planets move but it did not explain why the planets move in the way that they do.

Clearly, Aristotle's age-old explanations did not work, but what should they be replaced with? Kepler's best guess was that magnetism was somehow involved. He imagined that the rotation of the Sun produced a kind of magnetic whirlpool in space that emanated outwards and that this swept the planets around the Sun. Of course, these speculations were completely wrong, and the correct explanation would be provided by Isaac Newton, as we will see in Chapter 4. But there was another great clue that Kepler discovered that would be critical for later scientists such as Newton. To get an idea about the significance of this discovery, we will take a look at the pendulum, as Horrocks and other seventeenth-century scientists did, to see what insight it offers for the motion of the planets.

Galileo discussed the motion of a pendulum swinging in a single plane, like the pendulum in a grandfather clock, and this is the image that we usually have of a pendulum. However, a pendulum bob that is freely suspended from a string can simultaneously swing in two perpendicular directions, that we might define as *left-to-right* and *back-and-forth*, for instance. This more general pendulum is known as a spherical pendulum. The left-to-right and back-and-forth motions combine to form a circuit that is traced out by the bob.

What is the shape of this circuit? Well, because the period of the swing is independent of its size, the left-to-right motion must take the same period of time as the back-and-forth motion. In other words, the swings in the two perpendicular planes will remain in step. The motion will therefore be periodic, and the shape traced out by the bob will close on itself and repeat indefinitely. If the swings in the two perpendicular directions are equal in size, the shape is a circle. If one swing is bigger than the other, as will generally be the case, then the shape is a squashed circle—or, in other words, an ellipse.

> **Puzzle 5** Imagine that the force of gravity operates like a conical
> pendulum. When the radius of the orbit of our pendulum is 10 cen-
> timetres, it takes four seconds to complete one orbit. If the radius
> of the orbit is increased to 40 centimetres, how long does it take to
> complete one orbit?
>
> HINT: Consider Galileo's discovery about pendulums.

This looks like a promising start for a model of a planetary orbit. It certainly offers something to work with that is readily to
hand and easier to play with than the planetary system itself. For
these reasons, Jeremiah Horrocks made spherical pendulums and
explored their properties (Figure 3.11).

There is one immediate difference between the pendulum and
the planet, which is that the force acting on the pendulum bob is
towards the centre of each of the two perpendicular swings. So,
whereas in the planetary case the force is towards the Sun which
is located at a *focus* of the planet's orbital ellipse, the force on the
pendulum bob is towards the *centre* of the ellipse.

The second key difference between a spherical pendulum and
the orbit of a planet shows why Kepler's Third Law was so important for later researchers. Galileo's observation in the cathedral
tells us something important about the rate at which the bob

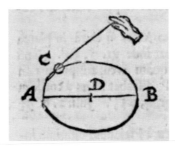

Figure 3.11 A diagram from the posthumous papers of Jeremiah Horrocks, in which he illustrates the use of a conical pendulum to mimic
gravity.

completes a circuit. Because the period of a swing is independent of the amplitude of the swing, the time taken for the bob to complete its elliptical orbit must be independent of the size of the orbit. This means that if the planets were held in the solar system by a similar force, each one would take the same period of time to orbit the Sun. Mercury, Venus, Mars, Jupiter, and Saturn would all complete their orbit in one year, just as the Earth does. This is not what we see at all—it takes Saturn almost thirty years to orbit the Sun.

Answer to Puzzle 5 The orbital period of the pendulum bob is independent of the radius of the orbit, so the bob will still take four seconds to complete one orbit, even though the orbit is four times as large. For a pendulum, the equivalent of Kepler's Third Law is that the period of the orbit is independent of the size of the orbit. However, this is not how the planets orbit the Sun. It proves that the force of gravity does not operate like the force that controls the motion of a pendulum bob. Kepler's Third Law would be the key to finding the correct force law obeyed by gravity.

From the way in which the period changes with the size of the orbit, it is possible to determine how the force changes with distance. In the case of the pendulum, the periods of the orbits are independent of the size of the orbits. This implies that the force acting on the bob is proportional to the distance from the centre. At twice the distance, the force is twice as large, at three times the distance it is three times as large, and so on. This makes sense in the case of the pendulum bob, because it means that the further the bob is from the centre, the stronger the force pulling it back to the centre and therefore the faster the bob moves with the result that it completes its bigger orbit in the same time.

This is very different to the way that the solar system works. In the solar system, the further a planet is from the Sun, the slower it travels. It is clear that the relationship between the period of an orbit and the size of the orbit is a good clue to the force that is acting, and this is just the sort of relationship that Kepler spent his whole life searching for. As we saw in the previous chapter, Kepler found just such a relationship in the heavens. What Kepler discovered was that the square of the period of a planet's orbit is proportional to the cube of the radius of the planet's orbit. And this relationship is the key to the force that is acting on the planet. Turning the key would unlock the whole of physics and bring about the most sensational revolution in the history of science. The person who turned the key is the hero of our next chapter.

The Goddess of Love

Kepler had predicted, in the Rudolphine Tables, that Venus would cross the face of the Sun in 1631. Such an event had never been seen. A number of Europe's astronomers watched out for the transit but with no success. We now know that it was not visible in Europe, as the Sun had set before the transit began; it was only visible from the opposite hemisphere of the Earth where no one was watching.

By 1639, Horrocks was living in Much Hoole, a tiny village midway between the Lancashire coastal town of Southport and the market town of Preston. Horrocks knew that, according to the Rudolphine Tables, in November of that year there would be a near miss and Venus would pass just below the Sun's disc. Throughout October, Horrocks tracked the position of Venus from evening to evening. Then, with just over a month to go, he suddenly realized that Kepler was mistaken. Venus was on course to pass directly across the face of the Sun. This was remarkable, as transits of Venus are very rare events; we know now that every 120 years or so a pair of transits will occur, separated by an interval of eight years. These two transits are visible from opposite hemispheres of the Earth. All earlier transits visible from Europe would have occurred before the invention of the telescope, so the young Horrocks, isolated in an obscure village in a remote area of Lancashire, had the opportunity to see an astronomical event that had never been seen before. On 26 October 1639, he excitedly penned a letter to his friend William Crabtree, a draper who lived in Broughton, just outside Manchester:

> *My reason for now writing is to advise you of a remarkable conjunction of the sun and Venus on the 24th November when there will be a transit. As such a thing has not happened for many years past, and will not occur again in this century, I earnestly entreat you to watch attentively with your telescope, in order to observe it as well as you can.*[16]

In preparation for the great event, Horrocks took a sheet of card and drew a six-inch circle, on which he intended to plot the course of Venus across the face of the Sun. He inscribed the circumference with 360° marks, and drew a diameter, which he divided into 120 parts (Figure 3.12).

His calculations suggested that the transit would begin at 3 p.m. on 24 November. But to ensure that he would not miss this once-in-a-lifetime event due to a misplaced confidence in his own observations, Horrocks set up his telescope a day early and projected the image of the Sun onto the back wall of his room such that the Sun exactly filled the 6-inch circle that he had drawn. He could see one small sunspot, but nothing un-usual. The following day, he continued with his vigil. Periodically returning from his duties he kept watch on the projected image of the Sun as it appeared and disappeared behind the clouds. Still there was nothing. Time ran on towards mid-afternoon and the hour when he expected the transit to start. Then, he was sud-denly 'called away by business of the highest importance, which with propriety could not be neglected'.[17] When he returned at a quarter past three, the clouds had dispersed and he saw, in his words: 'a most agreeable spectacle'—the shadow of a perfectly spherical body that had already fully entered onto the face of the

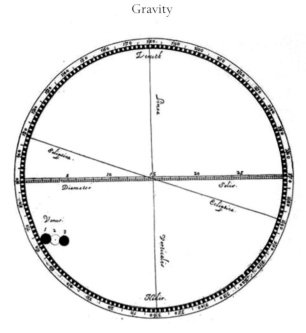

Figure 3.12 The diagram on which Horrocks marked the passage of Venus across the Sun, as published by Hevelius.

Sun. He watched eagerly for the next half an hour until sunset, regularly recording the position of Venus as the transit progressed (Figure 3.13).

Meanwhile, in Broughton, William Crabtree had made his own preparations for viewing the transit. Over Manchester, the skies were covered with thick cloud, and the disconsolate Crabtree thought that his opportunity to see this rarest of events would be lost to the Manchester weather. Then, suddenly, at just after half past three, the skies cleared, and a stunned Crabtree could see Venus on the face of the Sun for the quarter of an hour until sunset.

It is important to note that it is very dangerous to look directly at the Sun, with or without an optical instrument. Doing so can lead to permanently damaged eyesight.

Figure 3.13 The transit of 1639 observed by Horrocks by Eyre Crowe (1891).

Expanding the Universe

Horrocks recorded his account of the transit in a remarkable document: *Venus in Sole Visa* (*Venus in the Face of the Sun*). Horrocks gives a detailed scientific analysis of his observations and argues convincingly that there could be no doubt that he had indeed seen Venus crossing the Sun's disc and not some other astronomical phenomenon, such as a sunspot. Horrocks and Crabtree both estimated the size of the disc of Venus, and they agreed that it was much smaller than had been anticipated by earlier astronomers. Horrocks concluded that Venus must be much further away than had previously been assumed and, if Venus was further away, this implied that the entire solar system must be much larger than anyone had suspected.

In later centuries, transits of Venus were recognized as ideal opportunities to determine precisely the size of the solar system. By recording the time at which a transit begins at two well-separated locations, the parallax and, hence, the distance to Venus could be determined, and from this distance, the size of all the planetary orbits could be calculated. This was one of the main objectives of Captain Cook's first expedition to the Pacific; Cook recorded his observations of the transit of Venus from Tahiti on 3 June 1769.

In his writing, Horrocks implored astronomers to make their own telescopic observations and thereby acquaint themselves with the obvious inaccuracies in the existing astronomical tables. They could then confirm, to their own satisfaction, all his claims and conclusions. Horrocks had a rigorous and logical writing style. His account of the transit of Venus reads like a modern scientific paper in every respect but one. Horrocks was so moved by his discoveries that he interspersed poetry amidst his scientific arguments.

> *Divine the hand which to Urania's power*
> *Triumphant raised the trophy, which on man*
> *Hath first bestowed the wondrous tube by art*
> *Invented, and in noble bearing taught*
> *His mortal eyes to scan the furthest heavens.*[18]

An Incalculable Loss!

One year after observing the transit, Horrocks arranged to visit his friend William Crabtree in Manchester. On 4 January 1641, Crabtree awaited his friend, but Horrocks did not arrive. He was dead. No explanation of his sudden death has come to light. Crabtree, who recognized the enormous potential shown by Horrocks, was quite naturally distraught. He wrote:

> *Thus God sets an end to all earthly things. Ah departed friends (alas, the sadness of it all). O Horrocks most dear to me! Ah the bitter tears this has caused! What an incalculable loss!*[19]

He treasured Horrocks's papers and the memory of the young astronomer who had taken the study of the heavens beyond all his predecessors. Horrocks's work was known to a small group of Northern tradesmen and amateur astronomers but, over the next two decades, England was in turmoil and the writings of Horrocks remained unpublished. The Civil Wars were followed by the disruption of Cromwell and the Commonwealth. It was not until 1662 that Horrocks's account of the transit of Venus was finally published in Danzig. When the publication of this remarkable document was brought to the attention of the newly founded Royal Society, its members were severely embarrassed that such an important English work had not been recognized in England. The Royal Society gave the task of collecting and publishing all Horrocks's surviving papers to his friend from student days—John Wallis, by now the Savilian Professor of Geometry at Oxford University.

Horrocks was just twenty-two-years old when he died. The scale of his accomplishments in a few short years of isolated study is breathtaking. There is no way of knowing what he might have achieved had he lived. What is certain is that he was an extremely important stepping stone between Kepler and Newton, which means that his influence played some part in the creation of Newton's magnificent *Principia*, possibly the most important book ever written. Newton, who was not always generous with his praise for others, acknowledged the significance of Horrocks's theory of the Moon. Newton famously recalled that if he had seen further than others, it was by standing on the shoulders of giants.[20] One of these great figures was surely Jeremiah Horrocks.

During the Victorian era, there was a revival of interest in Jeremiah Horrocks, and a chapel dedicated to the remarkable young astronomer was built in St Michael's Church in Hoole. In the chapel, a marble tablet, and a stained glass window (Figure 3.14) commemorate the young astronomer who consolidated Kepler's revolution and laid the foundations for Newton's even greater revolution.

Figure 3.14 Stained glass window in St Michael's Church, Hoole showing Horrocks observing the transit of Venus.

4

Voyaging through Strange Seas of Thought

We are like dwarfs on the shoulders of giants, so that we can see more than they, and things at a greater distance, not by virtue of any sharpness of sight on our part, or any physical distinction, but because we are carried high and raised up by their giant size.
JOHN OF SALISBURY, *The Metalogicon* (1159) BK. 3, CH. 4

A Discussion over Coffee

Following a meeting of the Royal Society in January 1684, three of England's leading intellectuals—Christopher Wren (1632–1723), Robert Hooke (1635–1703), and Edmund Halley (1656–1742)—retired to a London coffee house to continue their discussions. All three men would be remembered for their contributions to science. On this evening their deliberations concerned the force that keeps the planets in orbit around the Sun. In particular, they were considering the laws of planetary motion derived by the German mathematician Johannes Kepler and what these laws implied about the force of gravity between the Sun and the planets.

In the early years of the century, Kepler had undertaken a painstaking analysis of the most accurate observations of the planets ever made. These had been compiled over the course of several decades by the astronomer Tycho Brahe. Kepler's conclusions are summarized in what became known as his laws of planetary motion which describe how the planets move around the Sun, as set out on page 77.

The three Fellows of the Royal Society could see that Kepler's Third Law, which relates the period of a planet's orbit to its distance from the Sun, might be the key to the problem as it seemed to suggest that gravity becomes weaker as the distance between the Sun and a planet grows. More precisely, they believed that it implied that the force of gravity declines as the inverse square of distance, which means that at twice the distance the force falls to a quarter of its original value, at three times the distance, it is a ninth of its original value and so on. But their proposal was quite tentative as they did not have the necessary mathematical tools to demonstrate this conclusively.

Wren realized that the case for the inverse square law would be clinched if it could be demonstrated that this force law would produce elliptical orbits as described by Kepler's First Law. So he offered a prize to anyone who could prove this relationship. Hooke claimed that he had a proof but was unable to produce it and no-one else could find a solution. So, in August 1684 Halley travelled to Cambridge to consult the Lucasian Professor of Mathematics, Isaac Newton, in his rooms in Trinity College.[1]

I Have Calculated It!

The mathematician Abraham De Moivre has left us an account of this momentous meeting:

> After they had been some time together, the Dr asked him what he thought the curve would be that would be described by the planets supposing the force of attraction towards the Sun to be reciprocal to the square of their distance from it. Sir Isaac replied immediately that it would be an ellipse. The Doctor, struck with joy and amazement, asked him how he knew it. Why, saith he, I have calculated it. Whereupon Dr Halley asked him for his calculation without any farther delay. Sir Isaac looked among his papers but could not find it, but he promised him to renew it and then to send it him.[2]

Newton duly wrote up his proof and sent it to Halley in November of that year. When Halley read the letter, he was astonished. It was clear that Newton had developed a whole system of mechanics but had told nobody. Halley realized the significance of Newton's discoveries and knew that they had to be published. He was determined to persuade Newton to write up a complete description of his system. Halley even agreed to pay for the publication and so Newton set about the task. It was a monumental undertaking that would result in a revolution in our understanding of the universe. The publication in 1687 of Newton's *Philosophiæ Naturalis Principia Mathematica* (*The Mathematical Principles of Natural Philosophy*), usually known simply as the *Principia*, would trigger the birth of modern science.

It is unlikely that Newton ever claimed the prize offered by Sir Christopher Wren, which was a book worth 40 shillings. But he did win undying fame with the publication of the *Principia*, a book whose value is beyond calculation.

Isaac Newton

How did Newton produce the answer to the question that had defeated all the other great minds of the seventeenth century?

Isaac Newton (1642–1727) was born on Christmas Day 1642 by the reckoning of the old Julian calendar that was still in use in England at the time. He was born in the hamlet of Woolsthorpe near Grantham in Lincolnshire. Newton's father had already died by the time he was born; he would soon also effectively be motherless, as his mother moved in with a new husband and left the baby Isaac to be raised by her parents. From this unpromising start in life, Newton would become one of the greatest mathematicians of all time and the greatest scientist who has ever lived. He would spark a scientific revolution that would transform the world and devise an approach to science that remains at the heart of engineering and the physical sciences today.

In 1661, Newton went to Cambridge University and entered Trinity College. At this time, Cambridge was something of an intellectual backwater, but Newton had access to the books of Galileo, Kepler, Descartes, and Gassendi. Unlike many of his predecessors, Newton was not satisfied with vague philosophical explanations of the workings of nature; he was a very practical man who combined a willingness and ability to perform experiments with a rigorous mathematical outlook on the world. In later life, Newton described the years 1665–1666 as his most productive. As he put it:

> For in those days I was in the prime of my age for invention & minded Mathematicks & Philosophy more than at any time since.[3]

This included a period when he returned to Woolsthorpe, as the university was closed due to an outbreak of the plague. During this period when Newton was in his early twenties, he invented calculus, which is one of the three most important inventions in the history of mathematics.[4] He also worked out many of the principles of mechanics that would eventually form the basis of the *Principia*.

Many years later, Newton told the story that it was during this time, when he was sitting in the garden of his family home, that the fall of an apple inspired him to think about gravity and ponder the possibility that the apple was attracted to the ground by the same force that holds the Moon in orbit around the Earth. Newton would illustrate this idea in the *Principia*. He asked his readers to imagine a cannon ball fired from a mountaintop. The cannon ball would travel along an arc until it hit the Earth's surface. If it was fired with a greater velocity it would disappear over the horizon but would eventually curve back to the ground some way around the Earth's circumference. Newton argued that if the cannonball was fired with sufficient velocity it would travel the whole way around the Earth before hitting the ground—it would be in Earth's orbit, just like the Moon (Figure 4.1).

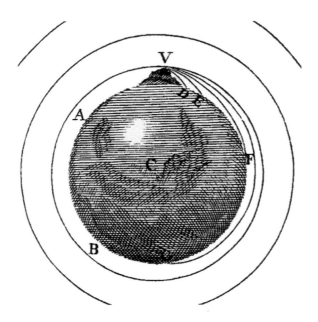

Figure 4.1 Diagram from Newton's *Principia* showing the trajectory of a projectile lauched from a mountain top. If the launch velocity of the projectile is sufficiently great it will enter Earth's orbit.

In 1957, 270 years after the publication of the *Principia*, the satellite Sputnik was launched into orbit by the USSR to become the first artificial satellite of the Earth. There are now many tens of thousands of satellites in orbit with a multitude of applications: television transmission, telephone communication, military surveillance, weather monitoring, and even Google Earth.

The Pattern of the Skies

Lo, for your gaze, the pattern of the skies!
What balance of the mass, what reckonings
Divine! Here ponder too the Laws which God,
Framing the universe, set not aside
But made the fixed foundations of his work.

The *Principia* opens with an ode penned by Halley, an excerpt of which appears above. Newton's language is altogether less flowery. Newton's style is a model of logic and rigour. He begins by setting down the fundamental principles of how objects move, with his three laws of motion:

Newton's First Law of Motion

All bodies move with a constant speed in a straight line unless acted upon by an external force.

This is the definition of natural or inertial motion. It is by no means obvious. Newton credited the philosopher and mathematician René Descartes (1596–1650) with the first correct statement of this law. The system of mechanics that Descartes derived from the law was, however, rather confused and certainly incorrect.[5] To an Earth-based observer, it might appear that unless a force is acting then a body will soon come to a halt. We all know that if we are pushing a heavy load, it will stop moving as soon as we stop pushing. A sailboat will stop when the wind ceases to blow. Even in situations where there is little friction—on an ice rink, for instance—we know that an object skimming over an icy surface, such as a curling stone, will eventually come to a halt. Everyday examples like these convinced Aristotle that a force is required to keep a body in motion and this misconception persisted for over 2000 years. In fact, in an environment such as outer space where there is no friction, objects will continue travelling in a straight line indefinitely if they are not subject to any force.

Next, Newton considered how forces act to alter the natural or inertial motion of an object. Newton was the first to give a precise meaning to the term *force*, which had previously been used in many different ways. Newton's critical insight was that force is directional. It is not sufficient to say that the application of a force will change the speed of an object, as Descartes thought.

What matters is that a force always acts on an object in a particular direction and the velocity of the object changes in the same direction. All this is encapsulated in Newton's Second Law, which defines what the effect of a force will be on the body to which it is applied.

Newton's Second Law of Motion

Force equals mass times acceleration.

This implies that the application of a force changes the velocity of a body, and this change is in the same direction as the force. Equivalently, the acceleration (or change in velocity) that a body undergoes when a force is applied is equal to the size of the force divided by the mass of the body, which means that the greater the mass of the body, the less its velocity will change due to the application of the force. This seems quite reasonable. We all know that it is easier to push a light object than a heavy one. A golfer could not propel a cannonball from the tee as easily as a golf ball. Newton's Second Law defines what is meant by a force and encapsulates the fact that a more massive body has greater inertia.

Galileo had recognized the role of inertia in the motion of projectiles moving close to the Earth's surface. Prior to Newton, however, no-one realized that the concept of inertia is also crucial to understanding how a force that is directed towards the Sun results in circular or elliptical motion around the Sun. Kepler thought that the force emanating from the Sun must sweep the planets around like a broom. Descartes imagined that space was filled with a system of cosmic whirlpools or vortices carrying the planets around in their circular motions. Other researchers believed that there must be at least two forces acting on a planet to keep it in its orbit—one directed towards the Sun and another pushing the planet around its orbit. Newton would provide an explanation based on the concept of inertia combined with a single force—gravity—that acts towards the Sun.

Newton's Third Law of Motion

To every action there is an equal and opposite reaction.

This law can be illustrated in many ways. Imagine two ice dancers; when they push against each other, they move off across the ice in opposite directions. It would be impossible for one ice skater to push against the other and race off, while the second ice skater remains stationary on the ice.[6] The shooting of a gun provides another example. When a gun is fired, the force on the bullet is exactly matched by the force on the gun which operates in the opposite direction, so as the bullet leaves the muzzle, there is inevitably a sharp kick. The bullet races off at a much higher speed than the gun recoils because the mass, and therefore the inertia, of the gun is much greater than that of the bullet.[7]

The Universal Law of Gravity

Others had considered gravity before Newton. Horrocks had recognized that the attraction between the Sun and planets was mirrored by similar attractions between other celestial bodies.[8] Hooke had argued that just as the Earth is bound together by its gravity, so is the Moon—which is why it is spherical.

What was completely new with Newton was the critical idea that gravity is universal. It acts between any two pieces of matter in the universe. This means that the total force between two objects is produced by summing the forces between all their constituent parts. This sounds horrendously complicated. Newton was the only person with the necessary mathematical skills to make any headway with such an idea. He proposed that the gravitational force between *any* two material particles decreases as the inverse square of the distance between the two particles. He expressed this as his universal gravitational law:

$$F = \frac{Gm_1 m_2}{r^2},$$

where F is the force of gravity, m_1 is the mass of the first particle, m_2 is the mass of the second particle, and r is the distance between them. G is a constant that determines the fundamental strength of gravity. It is known as Newton's constant.

Newton then proved the far from obvious fact that in the case of a spherically symmetrical mass, the total gravity of the mass is exactly the same as if the entire mass were located in a single point at its centre. This means that with regard to gravity, two massive spheres can be treated in exactly the same way as two massive point particles. This offers an enormous simplification when considering physical problems because many of the material objects of interest, such as stars and planets, are spherical to a very good approximation.

This means that the gravitational force between two celestial bodies can be expressed by exactly the same inverse square law as the universal law. For instance, the force between the Sun and the Earth is:

$$F = \frac{GMm}{r^2},$$

where now M is the mass of the Sun, m is the mass of the Earth, and r is the distance between the centre of the Sun and the centre of the Earth.

Gravity is intrinsically a very weak force, which is why it requires a huge accumulation of mass in an object like the Earth to produce any significant gravitational force. Mathematically, this is represented by the fact that G is small. Newton did not have a precise figure for the size of G as he did not know the mass of the Earth or of any other astronomical body.

Remarkably, this single force law coupled with Newton's three laws of motion is sufficient to explain the operation of the entire solar system and much else besides. This was the key to Newton's great breakthrough. The motion of the planets could now be understood from first principles. Newton (Figure 4.2) showed that, with this force law, each planet must move in an elliptical

Figure 4.2 Sir Isaac Newton in 1702 by Godfrey Kneller.

orbit around the Sun, just as Kepler had deduced from Tycho's
observations, and just as Newton had demonstrated in his letter to Halley.[9] The precision with which deductions made from
this small set of ideas matched observations would prove to be
extraordinary. Nothing like this had ever been possible before.
The *Principia* offered a quantitative and causal understanding of
the clockwork of the heavens.

A Visit to the Hollow Planet

So, as Newton showed, the gravitational field outside a spherical mass is described by an inverse square law. This is the cumulative effect of the gravity of all the matter forming the sphere. However, if large quantities of matter are arranged in a different way then the total gravitational force experienced by other material bodies may be quite different. The most dramatic instance of this is the gravitational field inside a hollow spherical shell of matter.[10]

The American science-fiction writer Edgar Rice Burroughs is most famous for his Tarzan stories. Burroughs also wrote a series of novels set in the land of Pellucidar which is located within the Earth. In these stories the Earth is imagined to be a hollow shell about 500 kilometres thick and the inner world is reached by airship through a passage at the North Pole.

> **Puzzle 6** What gravitational force would be felt within a hollow planet? Would the inhabitants be pulled toward the centre of the inner void? Would they be held to the inner surface? Or would gravity act on them in some other way?

Imagine taking a trip to the Hollow Planet. Prior to our arrival we might not even realize that the planet is hollow. Outside the planet, as it is spherical, the gravitational force produced by the planet will be the same as if all its mass were located in a point at its centre. The force will obey an inverse square law, just as it would for any other spherically symmetrical distribution of matter. When approaching the planet we might be surprised that its gravitational attraction appears to be very weak, but we would probably just assume that the planet is very low in the heavier elements. Perhaps, unlike the Earth, it might not have an iron core. But we would not be able to tell that the planet is actually hollow.

It would only be if we managed to make our way into the interior of the Hollow Planet like the visitors to Pellucidar that we would discover the reason for the planet's low mass. So what would we feel when we were on the inside within the planet? Would we stick to the inside of the shell and be able to walk around, as we do on the surface of the Earth? No: rather surprisingly, there would be no gravitational force whatsoever within the spherical shell of matter.

At the *exact centre* of a hollow shell of matter, we would expect the gravitational forces from the surrounding material to completely cancel out, whatever the force of attraction. The attraction of the material to one side is balanced by the attraction of an equal amount of material at the same distance on the other side. But when the force obeys an inverse square law, as it does in the case of gravity, we can move away from the centre of the hollow shell and the forces will still balance to leave no overall force (Figure 4.3). Gravity is the sum of forces from each piece of

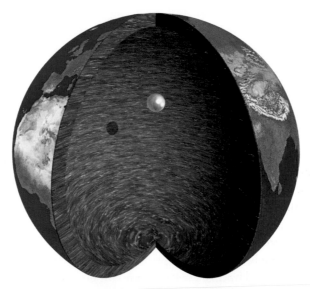

Figure 4.3 Within a hollow planet there would be no gravity. Any freely moving object within the planet would travel at a constant velocity, as no force would be acting on it.

matter and, when these forces diminish as the inverse square of distance, the force from each region on the near side of the shell exactly balances the force from a region on the opposite far side of the shell.

Answer to Puzzle 6 The force of gravity completely cancels within a spherical shell of matter, so there would be no gravitational force at all.

Journey to the Centre of the Earth

Now, the Earth certainly isn't hollow, but we can imagine taking another science-fiction journey, towards the centre of the Earth in the footsteps of the characters of Jules Verne. If we were to build a tunnel through the centre of the Earth, how would gravity vary as we pass through the tunnel? For the purposes of answering this question, we can consider the material from which the Earth is formed to be divided into two parts. When we are beneath the surface, there is a spherical shell of material that is further from the centre of the Earth than we are. There is also a ball of material that is closer to the centre than we are. As we have seen, the spherical shell above us will have no gravitational effect on us. The gravitational attraction on us will be due solely to the sphere of material beneath us (Figure 4.4). The strength of the force will be the same as if all this mass were located at the centre of the sphere. As we approach the centre the sphere beneath us will shrink in size. This means that as we fall from the surface of the Earth, the force of gravity diminishes gradually as we get closer to the centre. At the centre, we are within the entire Earth and the force of gravity will have completely disappeared.[11]

How will this be reflected in our journey through the tunnel? In freefall through the tunnel, we will be accelerated downwards towards the centre where we will reach our maximum velocity. After passing the centre, we will be decelerated, and we will come

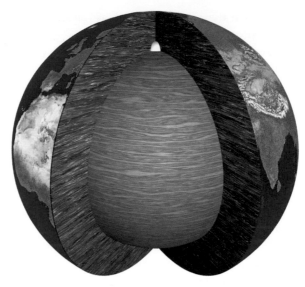

Figure 4.4 The gravitational attraction of the shell of material above the white capsule (shown cut away) cancels completely, so the white capsule is only attracted by the sphere of material beneath its position within the Earth (shown as the inner sphere).

to a halt as we reach the surface on the far side of the Earth. If we are unhindered, we will oscillate back and forth from one side of the Earth to the other just like a giant pendulum bob.[12] If such a tunnel were drilled through the Earth, how long would it take to reach the antipodes? We will seal the tunnel and remove all the air so that we are travelling through a vacuum. We step into the cable-free antipodes elevator, strap ourselves into our seats, then the robot driver takes off the handbrake and the elevator plummets like a stone into the shaft. Forty minutes later, we have arrived. We come to a gentle halt and the robot driver puts the handbrake back on—just in time, as a moment later we would have been racing on our way back home. Remarkably, we have travelled around 12,000 kilometres in under an hour and we haven't felt a thing; we have been weightless all the way. As we passed through the centre of the Earth, we were travelling at around 8 kilometres per second, and it did get rather hot.

The Moon and the Tides

Once upon a perfect night, unclouded and still, there came the face of a pale and beautiful lady. The tresses of her hair reached out to make the constellations, and the dewy vapours of her gown fell soft upon the land. This lady, whom all mortals call the Moon, danced a merry dance in the pathless sky, for she had fallen in love, and the object of her devotion was the Sun.

Kit Williams, *Masquerade* (1979)

Kepler believed that the Moon somehow causes the oceans to flow, which creates the tides. But Galileo ridiculed this idea and accused him of invoking occult influences to explain physical phenomena. Galileo had his own theory, in which the tides were produced by the oceans sloshing around as the Earth rotates. This was one of Galileo's main arguments for the Earth's rotation, but unfortunately it is completely wrong. One of Newton's most celebrated results in the *Principia* was his definitive explanation of the tides. The gravitational force between two objects decreases with distance, and this is the critical fact that enabled Newton to account for the tides. It means that the pull of the Moon is greater on the side of the Earth facing the Moon, than it is at the centre of the Earth, and the pull on the centre of the Earth is greater than the pull on the side of the Earth facing away from the Moon. If the Earth were completely fluid then the difference in the pull of the Moon on the different parts of the Earth would stretch the Earth in the direction of the Moon.

Now, the Earth is not completely fluid, of course, but the oceans are. The solid Earth is distorted by the pull of the Moon but not enough for us to notice, whereas the oceans flow. On the side of the Earth facing the Moon the oceans flow because the pull of the Moon on them is greater (because they are closer to the Moon) than the average pull on the bulk of the solid Earth. On the opposite side of the Earth, the oceans flow because they are pulled less by the Moon than the bulk of the solid Earth, because they are further away from the Moon (Figure 4.5).

Figure 4.5 This diagram is a schematic illustration of the forces acting on the Earth due to the gravitational pull of the Moon. The large arrows show the Newtonian force, which decreases with distance from the Moon. Forces 4 and 5 represent the net forces acting on opposite sides of the Earth. Force 4 is equal to force 1 minus force 2. Force 5 is equal to force 3 minus force 2 (force 5 points in the opposite direction to the other forces, because force 2 is bigger than force 3). These differential forces are known as tidal forces. They cause the tides to rise and fall.

The Earth rotates once every twenty-four hours. As it rotates, the region of the Earth facing the Moon moves, and so the positions of the bulges facing towards and away from the Moon change. Twice every twenty-four hours, we will be in the vicinity of a bulge. As the oceans are so much more fluid than the land, we notice this as a high tide, which occurs once every twelve hours or so.[13] (Clearly, there will also be a low-tide that occurs every twelve hours, midway between the high tides.)[14] For a more precise understanding of the details about the tides, there are a few more facts that need to be taken into account, but they do not alter the fundamental principles that are at work. The first thing to note is that although the Moon is the most significant body with regard to the tides, the Sun is also important, even though it is much further away. This means that the shape of the Earth and, more importantly, that of the oceans, are also distorted by the pull of the Sun. The result is that the tides are greatest when the Sun and Moon are aligned in the same direction, so that the distortions from the two bodies add up. The Earth, Sun, and Moon are in a line at both New Moon and Full Moon, and this is when the highest tides occur. The tides are smallest when the Sun and Moon are in perpendicular directions when viewed from Earth. These are the times when we can see half the face of the Moon (known to astronomers for dubious reasons as *First Quarter* and *Last Quarter*). At these times, the distortions that they produce in the shape of the Earth and its oceans partially cancel out.

To determine the exact timing of the tides at a particular place a bit more information is required. The Earth is spinning and the oceans are dragged around with it. This means that the bulge in the ocean due to the Moon's pull is constantly being carried around as the Earth spins and the oceans must continually flow back towards the position beneath the Moon, which takes time. So there is a small time lag between the expected time for high tide due to the position of the Moon and when it actually occurs. Over geological epochs, it has also had a rather dramatic effect on the Earth–Moon system. When the Moon formed, early in the Earth's history, the Earth would have been spinning much faster (a day might have been as short as about three hours), and the Moon would also have been much closer than it is now. Consequently, the tides were much bigger in the distant past and they produced a drag on the Earth's rotation, gradually slowing it down. Simultaneously the Moon gradually drifted further away from the Earth. This effect is much weaker now, but the tides are still slowing the Earth's spin and the Moon is still receding.

Britannia Rules the Waves

There is no drop of water in the ocean, not even in the deepest parts of the abyss, that does not know and respond to the mysterious forces that create the tide.
Rachel Carson, *The Sea Around Us*

The other factors that affect the timing and size of the tides are geographical. For instance, the shape of the Bristol Channel funnels the tides into the Severn Estuary and this produces the Severn Bore and some of the highest tides in the world.

Calculating the precise time and size of the tides at a particular place is complicated. Among other things it depends on predicting the position of the Moon, which is not easy. The Moon feels a strong gravitational attraction towards both the Earth and the Sun and this makes its motion difficult to analyse. To a good approximation the Moon's orbit is elliptical, but the axis of the ellipse changes direction, as Jeremiah Horrocks realized even before Newton's time. There are other important periodic variations in the orbit, such as the size of the eccentricity of the orbit and the degree of tilt of the orbit relative to the Earth's equator. Although each of these variations is regular, they all change at different rates, so the overall trajectory of the Moon does not repeat itself in a regular way. Before the age of the computer, this made any calculation of the position of the Moon and the corresponding prediction of the tides an extremely lengthy and arduous undertaking.

The Victorian scientist William Thomson (1824–1907) is better known today as Lord Kelvin. In 1872, Kelvin realized that it would be possible to construct a mechanical device that would perform the calculations by adding all the periodic variations together to give the height of the tide at any particular time and anywhere throughout the world. Within a year, Kelvin had completed his first ingenious design for such a machine. The machine contains dials that are used to set the oceanographic data for a particular harbour and the astronomical data for the starting time. It consists of several pulleys attached to assemblies whose gearing is set to sum the various contributions to the tides in the appropriate proportions. When the handle is cranked, the gears are driven and the height of the tides is computed and plotted on a roll of paper. In around four hours, the machine will calculate the tides for a full year.[15] Kelvin's original tide predicting machine is shown in Figure 4.6.

Figure 4.6 Kelvin's first tide predicting machine of 1872.

Kelvin's machine was the basis for tide calculators that were used for the following century, until the advent of digital computers. These machines produced extremely valuable information for a maritime nation such as Britain with its overseas empire. A single machine was used to calculate tides for the whole of India.

Mortals Rejoice!

Towards the end of his life Sir Isaac Newton mused:

> *I do not know what I may appear to the world, but to myself I seem to have been only like a boy playing on the seashore and diverting myself in now and then finding a smoother pebble or a prettier shell than ordinary, whilst the great ocean of truth lay all undiscovered before me.*[16]

On Newton's tomb in Westminster Abbey there is a Latin inscription which translates as:

> *Here is buried Isaac Newton, Knight, who by a strength of mind almost divine, and mathematical principles peculiarly his own, explored the course and figures of the planets, the paths of comets, the tides of the sea, the dissimilarities in rays of light, and, what no other scholar has previously imagined, the properties of the colours thus produced. Diligent, sagacious and faithful, in his expositions of nature, antiquity and the holy Scriptures, he vindicated by his philosophy the majesty of God mighty and good, and expressed the simplicity of the Gospel in his manners. Mortals rejoice that there has existed such and so great an ornament of the human race! He was born on 25th December 1642, and died on 20th March 1726.*[17]

Newton's mastery of the universe had a profound impact on the psychology of the British.[18] In the century following the publication of the *Principia*, Newton's discoveries were popularized in numerous books. They were presented in lectures and discussed widely.[19] The new science was made available to the whole of society, in books such as Voltaire's *The Elements of Sir Isaac Newton's Philosophy* (1738) and Francesco Algarotti's *Newtonianism for Ladies* (1737). It became generally accepted that a scientific understanding of the laws of nature was not only possible, but extremely useful in developing new technologies that could generate huge wealth.

The publication of mathematics textbooks increased to meet the demand for access to the new science. Engineers such as John Smeaton (1724–1792) undertook experiments to test the efficiency of mechanical devices. In 1759, Smeaton published *An Experimental Enquiry Concerning the Natural Powers of Water and Wind to Turn Mills and Other Machines Depending on Circular Motion*, in which he compared the efficiency of various types of waterwheel. This was the world in which engineers such as James Brindley (1716–1772), Matthew Boulton (1728–1809), James Watt (1736–1819), and Thomas Telford (1757–1834) grew up. Within a few generations, steam engines would be powering Britain's factories and the nation would be opened up by major engineering projects. Newton and his contemporaries at the Royal Society developed an understanding of science that would transform Britain's into the world's first great industrial power.

Alongside the early stirrings of British industry, Newton's ideas would be applied throughout the physical sciences. For over 200 years Newton would reign supreme.

Celestial Mechanics

If this is the best of possible worlds, what then are the others?

Voltaire, *Candide*[20]

In 1755, Lisbon, the mighty capital of the Portuguese Empire, was demolished by a catastrophic shockwave (Figure 4.7). The magnitude of the earthquake was about 9.0 on the Richter Scale, making it one of the biggest seismic events in recorded history. Disaster struck on the morning of All Saints Day, 1 November, when most of the population of 200,000 were attending church. Tens of thousands died in the earthquake and the subsequent tsunami; it has been estimated that a fifth of the population may have perished. This devastation prompted the philosopher

Figure 4.7 Earthquake at Lisbon, 1755.

François-Marie Arouet (1694–1778), better known as Voltaire, to write his most famous work, *Candide*.

John Michell (1724–1793) was also deeply affected by the Portuguese disaster and turned his attention towards the physical mechanisms behind earthquakes. Although Michell is now almost forgotten, he was one of the leading scientists of his day. He taught at Cambridge University and made many important scientific discoveries and he deserves to be much more well-known.[21] Following the Portuguese earthquake, Michell suggested that earthquakes spread out as waves through the solid Earth and are related to faults in geological strata. His book on the subject established seismology as a modern science, and Michell was rewarded by his election to the Fellowship of the Royal Society in 1760. In 1767, at the age of 42, Michell was appointed rector to the village of Thornhill, near Dewsbury in West Yorkshire, where he continued his scientific investigations.

Michell constructed his own telescopes, the largest being an impressive 10-foot instrument with a 30-inch mirror. But it is his ideas about astronomy that are most significant. Michell was the first to apply statistics to astronomy. He showed that if stars were scattered randomly throughout the sky, then it was extremely unlikely that we would see clusters of stars such as the familiar Pleiades or Seven Sisters. Michell concluded that star clusters must be genuine clumps of stars that are gravitationally bound together. Michell also demonstrated that there are far too many pairs of stars in the sky for them all to be due to chance alignments; he deduced that most such pairs must be orbiting each other in binary star systems.[22]

These enquiries led to even more astonishing speculations.[23] Michell considered the effects of gravity beyond the solar system and showed how the mass of a star in a binary system could be determined by studying the motion of its companion star. This analysis led him to speculate that there might exist stars that are invisible because light cannot escape from them.

When a ball is thrown upwards, its speed gradually decreases until its upwards flight ceases, and it falls back towards the Earth. If a projectile is hurled upwards fast enough, however, it may escape the Earth's gravity altogether and disappear into space. The escape velocity of the Earth is around 11 kilometres per second, while the escape velocity of the Sun is 600 kilometres per second. This is the speed that a missile or other projectile must reach in order to be released from the Earth's or the Sun's clutches. Michell considered a remarkable question: how massive must an object be if its escape velocity is to exceed the speed of light? He concluded that we would not be aware of any star with the density of the Sun but 500 times its diameter, as *its light would not reach us*. Michell then proposed that it might be possible to detect such *dark stars* in binary systems through the motion of their companion stars. We now know that a dark star such as Michell was imagining 250 years ago would collapse under its own gravity. It would be a black hole. We will soon be taking a much closer look at these mysterious objects.

Michell's final project was to determine the intrinsic strength of gravity. This might sound simple; after all, we measure the force of gravity every time we weigh ourselves. We cannot really answer the question in this way, however, unless we know the mass of the Earth. Michell designed an experiment that would measure the gravitational force not between the Earth and another object but between two solid metal balls. The mass of each ball could easily be determined, so by measuring the gravitational attraction between them, Michell could work out the value of the constant G in Newton's force law. This would also, indirectly, give him the mass of the Earth. Michell constructed the apparatus for the measurement but, unfortunately, in 1793, he died before he could perform the experiment. After Michell's death his apparatus was given to another leading scientist Henry Cavendish who would complete the experiment.[24]

The Silent Man

Henry Cavendish (1731–1810) was the son of Lord Charles Cavendish. He was an aristocrat and one of the richest men in England, as well as a leading scientific figure of the eighteenth century. Although he was totally dedicated to science, he was also excruciatingly shy and found conversation agonizing. According to one contemporary, he uttered fewer words in the course of his life than any man who lived to fourscore years, not at all excluding the monks of La Trappe'.[25]

Cavendish performed his experiments in his house in central London, where he communicated with his servants through written messages. He once happened to bump into one of his cleaners on his staircase and was so flustered by the experience that he had another private staircase built in his house so that he should never suffer this ordeal again. Despite his ill-ease with company, he rarely missed the weekly dinner of the Royal Society at the Crown and Anchor on the Strand. Cavendish was a meticulous experimenter who made significant advances in chemistry and physics decades before others who would receive the acclaim. Although Cavendish was highly respected by his colleagues, he published very little, so he did not always receive the credit he deserved for his scientific enquiries. Many of his discoveries were only revealed a century later, when James Clerk Maxwell (1831–1879) studied his notebooks.

By the time Cavendish obtained the apparatus built by Michell to determine the strength of gravity, its wooden frame had warped, so Cavendish had to have it substantially rebuilt. He also made other modifications to the equipment before performing the measurements that Michell had devised. Cavendish gave Michell full credit for its design, and it is usually known today as the Michell–Cavendish Experiment. The measurements were completed in 1798. Cavendish quoted his results as a determination of the average density of the Earth, which he revealed to be 5.48 times the density of water, which is within 1% of the well-established modern figure. This was an incredible achievement for such a difficult experiment, performed in Cavendish's front room over two centuries ago. From his result, it is straightforward to calculate the value of Newton's constant G, which encapsulates the fundamental strength of the force of gravity.

Cavendish died in 1810. Half a century later his relative William Cavendish, the 7th Duke of Devonshire, gave an endowment to the University of Cambridge in honour of his scientific research. This money was used to build the university's physics laboratory. It was named the Cavendish Laboratory by James Clerk Maxwell, who would be the first Cavendish Professor. The Cavendish Laboratory would be the home of world-changing research, from the discovery of the electron to the structure of DNA.

The Music of the Spheres

William Herschel (1738–1822) was born into a musical family in Hanover on 15 November 1738 (Figure 4.8). William's father Isaac was a member of the regimental band of the Hanoverian Guards and as a youth, William joined him in this occupation. At this time the crown of Hanover was united with that of Britain under the reign of George II. In 1756, the Seven Years War broke out,

Figure 4.8 William Herschel by Lemuel Francis Abbott (1785).

France invaded Hanover, and the Hanoverian forces were over-whelmed at the Battle of Hastenbeck. Although William was not injured, he spent the night sheltering in a ditch—an experience that convinced him that he was not suited to the life of a soldier. Without waiting to be discharged from the Hanoverian forces, Herschel managed to find his way to England to build a new life as a musician.

By 1766, Herschel had obtained the position of organist at the Octagon Chapel in Bath. He also became the Director of Public Concerts in this fashionable spa town, composing and perform-ing many concertos and symphonies. Despite his obvious mu-sical talents, Herschel is not remembered today for his music.

In 1773, he took an interest in astronomy and started grinding lenses and building his own telescopes. With the help of his sister Caroline, Herschel would make systematic observations of a variety of astronomical objects, including the planets, double-star systems, and comets. Herschel became obsessed with perfecting his optical equipment. His passion was all-encompassing, and he would spend sixteen hours a day at the tedious job of grinding the lenses for his astronomical instruments. While he ground the glass, Caroline read to him and even spoon-fed him while he continued with the laborious task. But it was worth it. Herschel's telescopes proved to be better than any others in Britain, including those of the Astronomer Royal at the Greenwich Observatory.

In March 1781, while searching for double stars, Herschel came upon a new object that appeared as a disc in his telescope. This clearly could not be a star, as stars are so distant that even the biggest still appear point-like through a telescope. Herschel initially thought that it must be a comet, but it was a very unusual comet as it appeared to be circular and it did not have a tail. Herschel monitored the object over the next few nights and recorded that it was moving slowly against the background stars. From the rate of its motion, it was clear that it was more distant than Saturn.[26] Herschel wrote to a circle of leading astronomers announcing his discovery. Soon it became clear that the object within Herschel's sights was not a comet, it was a far more sensational discovery: a new planet, the first new planet to be discovered in recorded history. Herschel proposed that it should be called *Georgium sidus*, the Georgian star, after the Hanoverian monarch. Fortunately, this name soon fell out of favour. We now know the planet as Uranus, named after the Greek god who was father to Cronos (usually identified with the Roman god Saturn). Herschel went on to make many other discoveries including two satellites of Uranus. His sister Caroline was a great astronomer in her own right and is one of the greatest comet finders of all time.

Uranus is just about bright enough to be visible to the naked eye, so it is rather surprising that it was not discovered until over 170 years after the invention of the telescope. Astronomers soon realized that it had been observed on several occasions over the years, and even plotted on star maps by astronomers who had assumed that it was just another fixed star. The distant Uranus moves quite slowly against the background stars, completing a full orbit in eighty-four years, so these early sightings of the planet proved to be very useful for determining its precise orbit. William Herschel died in his 84th year in August 1822, living almost long enough for his planet to complete one full orbit of the Sun.

An Embarrassing Episode

By the early years of the nineteenth century, it was becoming clear that Uranus was not keeping to the path predicted by the astronomical tables. Every attempt to pin down an elliptical orbit for the planet seemed to work for a period of time but within a few years the planet would drift from its expected position. This became known as the Uranus problem. The issue was all the more glaring, as Newton's celestial mechanics was supposed to represent the perfection of the exact sciences. The planetary clockwork was understood in exquisite detail and performed exactly as predicted with regards to all the other planets, but Uranus would not behave.

In 1828, a series of observations of Uranus were taken under the supervision of George Biddell Airy (1801–1892), Cambridge Professor of Astronomy. They indicated a discrepancy of 12 seconds of arc when compared to the data in the best current astronomical tables. This is a small error—about 150th of the diameter of the Moon—but it was certainly significant when compared with the accuracy of the predictions for the other planets and was well beyond what would be expected from observational errors in the early part of the nineteenth century. By 1829, the error had grown to 23 seconds of arc, and the following year it reached 30 seconds of arc. Something was clearly wrong with the astronomical tables.

As Airy pointed out in his 1832 *Report on the Progress of Astronomy*, Uranus seemed to be unwilling to conform to an elliptical orbit. A number of unconvincing suggestions were put forward to explain the anomaly. Perhaps Uranus had collided with a comet and this had affected its trajectory. Perhaps Newtonian gravity needed to be modified over distances as large as that between the Sun and Uranus.

In 1841, a young undergraduate called John Couch Adams (1819–1892) came across Airy's report in a Cambridge bookshop and set out to solve the puzzle.[27] Adams could see at once that the most likely explanation was the existence of another planet orbiting the Sun beyond Uranus, and that this unknown planet was tugging on Uranus and causing the deviations in its orbit. This possibility had already been suggested by others, but no-one had tackled the problem mathematically and shown that such a solution would work. Given the existence of an extra planet with a known orbit, it would be relatively straightforward to calculate the effect of its gravitational pull on each of the other planets. But working in the other direction would present difficulties on a totally different scale.

Part of the problem was that a very massive planet orbiting at a great distance from Uranus would have a similar effect to a less massive planet orbiting close to Uranus. Furthermore, the new planet's orbit might be circular but, on the other hand, it might be a highly eccentric ellipse that would see the planet receding into the depths of space. It was certainly a problem that no-one had ever solved before. Indeed, no-one had even contemplated such a problem before. But Adams was a very capable mathematician, with exactly the right skills for the task.

Adams undertook the arduous calculations and by September 1845, he had computed the position of the hypothetical planet and compiled clear instructions for where astronomers should look to find it. He wrote up his findings and sent them to Airy who by now was director of the Royal Observatory at Greenwich. Airy was a meticulous man who ran his observatory like clockwork. He instructed his employees never to throw away any paper. Everything had to be documented before being stowed away for safe keeping. A colleague joked that, 'if Airy wiped his pen on a piece of blotting paper he would duly endorse the blotting paper with the date and particulars of its use, and file it away amongst his papers'.[28] In Airy's view the task set before the Observatory was clear. It was the precise measurement of the heavens in order to regulate the time and to aid navigation between Britain and her dominions. The observatory lay at the heart of the Empire with the express purpose of regulating its pulse and maintaining its smooth running and commerce. Naturally, there was no place for speculative searches after new discoveries. Airy received the letter from Adams and promptly filed it away.

Meanwhile, on the other side of the channel, the French mathematician Urbain Le Verrier (1811–1877) had reasoned in the same way as Adams and had made his own calculations of the position of an unseen planet that might be pulling Uranus out of line. Through the summer of 1846, Le Verrier made several attempts to persuade French astronomers to look for his planet, meeting with no more success than Adams. Finally, on 18 September, he wrote to Johann Gottfried Galle (1812–1910), an assistant at the Berlin Observatory, who received the letter five days later. The director of the observatory, Johann Franz Encke (1791–1865), was unimpressed by the suggestion of a new planet but Galle was keen to take up the challenge. Encke gave his permission and, along with his colleague Heinrich D'Arrest (1822–1875), Galle started the hunt that very night. The pair set out on a systematic search of the region of sky that Le Verrier had indicated. They compared the position of each star that they could see to an up-to-date map of the heavens. Within just half an hour they spotted a point of light that was not on the map—the planet Neptune had been found.

Figure 4.9 Cartoon from the 7 November 1846 issue of the French magazine *L'Illustration*, showing Adams looking for Neptune in the wrong direction before finding it in Le Verrier's notebook.

The following night, the excited Galle showed the planet to Encke. They confirmed that it had moved slightly since the previous night by the amount that would be expected for a planet at that distance, and that it appeared through the telescope as a small disc rather than a mere point of light. Encke wrote to Le Verrier: 'Allow me, Sir, to congratulate you most sincerely on the brilliant discovery with which you have enriched astronomy. Your name will be forever linked with the most striking proof imaginable of the validity of the Law of Universal Gravitation.'[29] The news soon reached London, where it was heralded as one of the most embarrassing episodes in the history of British astronomy. While Britain and France were vying to carve up the Earth, the French had stolen a whole new world from under the noses of the British (Figure 4.9).

The Search for Planet Vulcan

Spock: Check!
Kirk: Checkmate!
Spock: Your illogical approach to chess does have its advantages on occasion,
      Captain.
Kirk: I'd prefer to call it inspired.
Spock: As you wish.[30]

Following his triumph with the planet Neptune, Le Verrier drew up plans for an in-depth analysis of the entire solar system, taking into account the mutual gravitational influences between all the planets. In 1854, he was appointed director of the Paris Observatory, where he combined his theoretical analysis with an observational survey of the motion of the planets. After compiling the data, he was able to account for the orbital characteristics of each planet. Every orbit fitted his calculations perfectly with just a single exception—the planet Mercury—the messenger of the gods, racing around the Sun on the innermost path in the solar system.

Mercury has the most eccentric of all the planetary orbits, and observations showed that the orbit precesses. More precisely, the direction of the axis of Mercury's orbit rotates by 574 seconds of arc, or just under one-sixth of a degree, per century. In around 225,000 years, Mercury traces out a complete orbital rosette. Le Verrier calculated that the gravitational attraction of Venus, the planet that approaches Mercury most closely, has the biggest effect on Mercury's orbit. It accounts for a shift of 277 seconds of arc per century. The giant planet Jupiter adds another 153 arcseconds, the Earth chips in 90 arcseconds, and the rest of the planets add about 11 arcseconds per century. These contributions amount to a total of 531 arcseconds, which leaves an unaccounted for 43 seconds of arc.[31]

Le Verrier assumed that the wanderings of Mercury must have a similar cause to the wanderings of Uranus. Following his successful prediction of the existence of the planet Neptune, he proposed that there must be another unknown planet within the orbit of Mercury that was responsible for producing the unexplained shift in its position. Mercury is close to the Sun, which can make it quite difficult to see. Although it is very bright, it tends to be lost in the glare of the Sun as it is only visible in the twilight of sunset or sunrise. A planet within the orbit of Mercury would be even more difficult to spot, so Le Verrier's suggestion of another hidden planet could not be ruled out without further investigation. The hypothetical planet was named *Vulcan* after the metalworking smith of the gods baking in the heat of his forge.

Even though Vulcan would be very hard to find under normal circumstances, during a total eclipse it should be visible close to the Sun. Nineteenth-century astronomers were on the lookout and there were several reported sightings in the second half of the century. Le Verrier died in 1877, no doubt convinced that his proposal was correct, but none of the sightings could be confirmed, for the very good reason that Vulcan does not exist. It was not until 1915 that the puzzle of Mercury's wanderings was solved. The solution was remarkable and totally different to the explanation of the discrepancies in the orbit of Uranus. It involved a discovery that was even greater than the discovery of a new planet. The mystery would be solved by Albert Einstein.

In the 1780s, one century after the publication of Newton's *Principia*, William Wordsworth was a student at the college next to Newton's in Cambridge. Wordsworth's poem *The Prelude* looks back to his days at St John's College, ending with Wordsworth dreamily surveying the chapel of Trinity College from his rooms:

> *And from my pillow, looking forth by light*
> *Of Moon or favouring stars, I could behold*
> *The antechapel where the statue stood*
> *Of Newton with his prism and silent face,*
> *The marble index of a mind for ever*
> *Voyaging through strange seas of Thought, alone.*[32]

5

The Great Ocean of Truth

*Light thinks it travels faster than anything else but it is wrong.
No matter how fast light travels, it finds the darkness has always
got there first, and is waiting for it.*
TERRY PRATCHETT, *Reaper Man.* (1991)

In 1862, James Clerk Maxwell published a set of equations that tied electricity and magnetism together and united them in his theory of electromagnetism. The electric and magnetic forces would henceforth be seen as two faces of a single coin. Maxwell's theory was the mathematical expression of the results of many experiments performed by Michael Faraday (1791–1867) and other earlier researchers over the course of several decades in the first half of the nineteenth century.[1] One of Faraday's great ideas was the concept of the *field* as a map of forces throughout a region of space. There is an electric field around an electrically charged body (Figure 5.1, Left), for instance, and a magnetic field around a bar magnet (Figure 5.1, Right).

Faraday's most important conclusions were that a changing or moving electric field generates a magnetic field, and that a changing or moving magnetic field generates an electric field. For instance, an electric current is produced by the movement of electrically charged particles, usually electrons, and the flow of these electrically charged particles produces a magnetic field, so a magnetic field is always present around an electric wire when a current is flowing (Figure 5.2, Left). Conversely, moving a bar magnet near

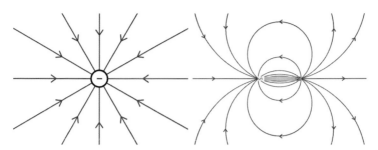

Figure 5.1 Left: The electric field around a negatively charged body such as an electron. The arrows indicate the direction of the force on a positively charged body. Right: The magnetic field around a bar magnet. The arrows indicate the direction of the force on a sample magnetic north pole at each point in space around the bar magnet.

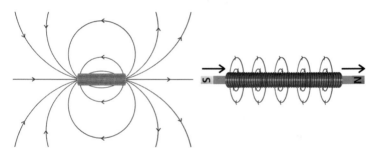

Figure 5.2 Left: The flow of an electrical current through a coil of wire generates a magnetic field (shown here in red). Right: A moving magnet generates an electric field (shown here in blue), that causes a current to flow in a coil of wire.

a wire generates an electric field, and this electric field causes the electrons in the wire to move and, thereby, produces an electric current (Figure 5.2, Right). The interplay between these two aspects of the electromagnetic force lies at the heart of the electricity industry and the electrical devices that we are so familiar

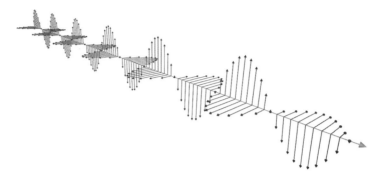

Figure 5.3 An electromagnetic wave. The blue arrows represent the electric field and the red arrows represent the perpendicular magnetic field.

with. Maxwell's equations captured the interrelationship between electricity and magnetism in a neat and concise fashion.

Maxwell's equations offered an added bonus. The equations could be combined into an equation describing the propagation of an electromagnetic wave—one formed of an electric field and a magnetic field oscillating in perpendicular directions. As the electric field changes, it generates the magnetic field and as the magnetic field changes, it generates the electric field in a mutually self-sustaining manner. The wave propagates in a direction perpendicular to the directions in which the electric and magnetic fields oscillate, as shown in Figure 5.3.

In one of Faraday's experiments, he passed a beam of polarized light through a magnetic field and showed that the magnetic field affects the polarization of light, which hinted at a previously unsuspected connection between light and electromagnetism. Faraday tentatively suggested such a link, but he was unable to prove it to his own satisfaction. When Maxwell worked out from his equations the speed at which his electromagnetic waves would be transmitted, it all fell into place; to within a small margin, the speed was equal to the most up-to-date measurements of the speed of light. Maxwell visited Faraday on his deathbed and told him that once more he had been correct, but Faraday was already too ill to understand the significance of his words.

In the second half of the 1880s, the German physicist Heinrich Hertz (1857–1894) provided a direct confirmation of Maxwell's identification of light as an electromagnetic wave. Hertz generated radio waves from a spark produced by passing an electric current across a spark gap, just like the spark plugs in a car engine which sometimes produce radio interference in this way. Hertz detected the radio waves on the opposite side of his laboratory. He measured the wavelength of the waves and determined their velocity to confirm that it was the same as the speed of light. Radio waves are identical to light, but with a much longer wavelength.

It was now clear that the light that we receive from the stars is electromagnetic in origin. But if light rays are simply electromagnetic waves, what is it that is waving? This question perplexed Victorian scientists. Sound travels through the air as pressure waves. It is formed by regular variations in the pressure of the air—a sequence of compressions and rarefactions. But space is a total vacuum; it is empty, so how could it transmit electromagnetic waves? What was the nature of the medium that undergoes the electromagnetic vibrations that we perceive as rays of light?

There was no easy answer. The standard response was that the whole of space must be permeated by a substance known as the *luminiferous ether*, and that electromagnetic waves were vibrations in this material. But this substance would need to have some very remarkable properties. It had to be sufficiently robust to propagate waves at the extremely high velocity of the speed of light. But equally, it had to be so insubstantial that it would offer no resistance to the motion of the planets so that they would continue in their orbits indefinitely, in accordance with Newtonian physics, and not just rapidly spiral into the Sun. There were some who argued rather unconvincingly that the ether was proof of the existence of God, as only God could make a substance with these impossible properties.

Making Waves in the Ether

What physicists needed was an experiment that would provide incontrovertible evidence of the existence of this extraordinary medium with its supposedly miraculous qualities. The ether was assumed to provide a stationary medium through which the Sun, planets, and stars were moving. So its existence could be unambiguously demonstrated by measuring the motion of the Earth through this background. And, if light was a vibration of the background ether, then it should be possible to detect the Earth's motion by measuring the speed of light in different directions. To settle the question, physicists required two measurements of the speed of light—one in the direction of Earth's motion and one in a perpendicular direction.

The American physicist Albert Michelson (1852–1931) worked for many years on techniques for accurately measuring the speed of light. From 1882 onwards, Michelson perfected a precision instrument known as an *interferometer* (Figure 5.4). It worked by splitting a beam of light and reflecting each half-beam numerous times between pairs of mirrors to increase their path length before recombining the beams to produce an interference pattern; bright fringes appear where the peaks from the two half-beams arrive together, and dark fringes, where a peak from one half-beam meets a trough from the other. In 1887, Michelson collaborated with Edward Morley (1838–1923) with the aim of detecting the Earth's motion through the ether with his interferometer. The assumption was that the Earth's motion through the stationary ether would create an *ether wind*, as it was termed, and light waves would be sped up or slowed down depending on whether they were moving with or against this wind, so that light travelling in the same direction as the Earth would travel at a different speed to

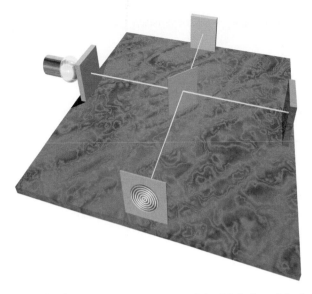

Figure 5.4 A schematic representation of the Michelson–Morley Experiment. A beam of light is shone through a half-silvered mirror which splits the beam into two perpendicular beams. These beams are reflected back through the half-silvered mirror and on to a screen where they form an interference pattern. One of the mirrors is fixed, the other is movable so that the path length can be varied. The whole apparatus can be rotated, so that the orientation of the beams changes with respect to the Earth's motion through space.

light travelling in a perpendicular direction, and this would show up as a slight displacement of the interference pattern.

Michelson and Morley assembled their apparatus on a large block of sandstone in the basement of a stone dormitory where it was isolated from vibrations and thermal variations. The interferometer was floated on a bath of mercury so that it could be rotated with minimal friction. As the device was turned, each arm would point in a direction parallel to the Earth's motion twice during every complete revolution, and the effect on the interference pattern could be monitored through an eyepiece.

The interferometer was so sensitive that the Earth's motion should easily have been seen as a shift in the position of the peaks and troughs. But, to the amazement of Michelson and Morley, they could not detect any shift at all.

Michelson was shocked by the findings and repeated the experiment several times over the next few years with increasing precision but always with the same null result. It has been suggested that he never fully recovered. Yet, he continued his work on precision optical instruments and is recognized as one of the greatest of all experimental physicists. In 1907, he became the first American to win the Nobel Prize in Physics. Michelson's techniques play a very important role in today's gravitational physics, as we will see in Chapter 8. The sensitivity of modern interferometers is astonishing.

Michelson's skill as an experimenter was widely known and the null result of the Michelson–Morley experiment was immediately accepted, even though it presented physicists with a disturbing conundrum. In subsequent years, a number of theorists attempted to account for the results, but their explanations always seemed rather contrived and unnatural. Then, in 1905, a patent clerk (technical expert third class) working in a patent office in the Swiss capital, Bern, wrote a paper that would supply the correct explanation of the experiment. This obscure patent clerk was Albert Einstein (Figure 5.5), and his paper 'On the Electrodynamics of Moving Bodies' offered the world a new system of mechanics known to physicists as *special relativity*.

The Birth of an Einstein

Einstein was born in the German city of Ulm on 14 March 1879. No-one could have foretold that, in the not-too-distant future, his name would quite deservedly be synonymous with genius. Einstein was smart—very smart. This might seem obvious, but Einstein is often portrayed as being slow at school. In fact,

Figure 5.5 Einstein, photographed when he was a patent clerk.

Einstein's school record was very good. Einstein was introduced to philosophy during his teenage years. He taught himself calculus and became deeply interested in the ideas of thinkers such as David Hume (1711–1776), Immanuel Kant (1724–1804), and Ernst Mach (1838–1916). He would develop into the most philosophical of all the great physicists.

As early as the age of sixteen, Einstein would muse about physics by imagining ideal situations, such as sitting on a beam of light. Einstein turned the analysis of such *thought experiments* into an art form. The concept of the thought experiment was not new with Einstein, but Einstein emphasized its central role in the development of his ideas. It was a laboratory in the mind where he could test his understanding of the physical world. In many respects, it is a refinement of a type of argument where a philosopher stretches an idea into unfamiliar territory and draws conclusions from the possibly absurd repercussions.[2] The novelty of Einstein's use of philosophical arguments was that he united them with a thorough understanding of physics, and this enabled him to give his thought experiments a critical place in the construction of new theories of physics. In Einstein's hands, the thought experiment became an incredibly powerful tool.

The Cosmic Speed Limit

Einstein was adamant that the Michelson–Morley experiment was not a major influence on him. Rather, it was the structure of Maxwell's equations that inspired him to develop a system of mechanics that was capable of describing motion at speeds approaching that of light. As we have seen, Maxwell's equations predict that electromagnetic waves travel at the speed of light, which was exactly why physicists were able to identify light as an electromagnetic wave. The equations give a precise figure for this speed, and no indication that the rate might vary in the manner expected by physicists such as Michelson and Morley. Einstein argued that we should accept the message that Maxwell's equations were sending us, even if this meant there was a flaw in our understanding of motion.

According to Einstein, Maxwell's equations imply that Newtonian mechanics is unable to describe motion at speeds approaching the speed of light. He therefore set out to construct a new system of mechanics that would reduce to Newtonian mechanics for velocities that were small compared to the speed of light but would also correctly account for motion at much higher speeds.

Einstein based the new mechanics on just two fundamental principles. The first of these is the idea of relativity. Although this principle gives its name to Einstein's new system of mechanics, it was not really new with Einstein. Newton's First Law of Motion states that unless acted on by an external force, a body will move in a straight line at a constant speed. This establishes a notion of natural motion. This law applies to the motion of all objects; when no force is acting, they all move in the same way, that is in a straight line at a constant speed. This implies that there is no absolute way to determine the speed at which an object is travelling. If we throw a ball into the air, we can define its speed relative to the ground. We know that the Earth is moving around the Sun so we could work out the speed of the ball relative to the Sun. We also know that the Sun and planets are moving around the galaxy, so we could work out the speed of the ball relative to the centre of the galaxy. Which of these speeds is the correct speed of the ball? They are all equally valid—although the speed relative to the ground gives far and away the most convenient description of the motion of the ball. There is no absolute velocity; velocity is a relationship between two objects, so the velocity of one object only makes sense when defined relative to another object. When applied to Newtonian mechanics, relativity is usually referred to as *Galilean relativity*, since it is closely related to the concept of inertia, which was first understood by Galileo.[3]

Following in Galileo's footsteps, Einstein rested his first principle on the assumption that there is no experimental way to determine absolute velocity. In other words, whatever the velocity of our laboratory, as long as the velocity does not change, the outcome of any fundamental physics experiment will be unaffected; we cannot distinguish between different constant velocities. It is impossible to define a state of absolute rest, as Einstein pointed out at the start of his first paper on special relativity:

> The introduction of a 'luminiferous ether' will prove to be superfluous inasmuch as the view here to be developed will not require an 'absolutely stationary space' provided with special properties.[4]

Einstein's second principle *was* completely new. This principle states that there is a maximum possible speed for the transmission of any interaction and this speed is the speed of light in a vacuum. With this second principle, Einstein sliced through the Gordian knot of Michelson and Morley's experiment. According to Einstein, the Michelson–Morley experiment could never detect the Earth's motion by measuring the speed of light, because light would propagate at exactly the same speed, irrespective of the Earth's motion. Einstein summarized the two building blocks of special relativity in this way:

> *We will raise this conjecture (the purport of which will hereafter be called the 'Principle of Relativity') to the status of a postulate, and also introduce another postulate, which is only apparently irreconcilable with the former, namely, that light is always propagated in empty space with a definite velocity c which is independent of the state of motion of the emitting body.*[5]

Einstein's principle of relativity had disposed of the notion of absolute space. Strict adherence to his second principle, the invariance of the speed of light in a vacuum, meant that something else would have to give, and this would prove to be even more disturbing and counter to our everyday assumptions about the world. Accepting the speed of light as the cosmic speed limit implied that it was impossible to show unambiguously that two events are simultaneous. Information about the events would be transmitted at the speed of light. Although the two events might appear simultaneous to one observer, they would not be simultaneous to a second observer. For instance, if there were two supernovae in our galaxy, separated by a distance of 1000 light years, a civilization situated midway between the two supernovae might witness them simultaneously, whereas a second civilization in a different part of the galaxy might observe one supernova explosion 1000 years before the other.

Puzzle 7

Figure 5.6 Spiral galaxy showing the positions of supernova A, supernova B, and observer V, which form an equilateral triangle.

As shown in Figure 5.6, the positions of supernova A, supernova B, and the observer V form an equilateral triangle in the plane of the galaxy, so that both supernovae, which are 1000 light years distant, are seen simultaneously by the observer. A second observer sees supernova A 1000 years after supernova B. Where is the planetary system of the second observer located?

Answer to Puzzle 7 A second observer in a planetary system anywhere beyond supernova B, on the straight line extended from

continued

Continued

supernova A to supernova B, will see supernova A 1000 years after supernova B. Figure 5.7 shows an example of such a system. Planetary system C is 1000 light years beyond supernova B, so the light from supernova B takes 1000 years to reach this point, whereas the light from supernova A takes 2000 years to reach this point.

Figure 5.7 Spiral galaxy showing the positions of supernova A, supernova B, observer V, and observer C.

We might not be too concerned about giving up the notion of simultaneity but, as Einstein went on to point out, it affects our whole understanding of the nature of time.

We have to take into account that all our judgments in which time plays a part, are judgments of simultaneous events. If, for instance, I say that: 'A particular train arrives here at 7 o'clock,' I mean something like this: 'The pointing of the small hand of my watch to 7 and the arrival of the train are simultaneous events.'[6]

The only reasonable conclusion, according to Einstein, was that there could be no such thing as absolute time. In other words, the passage of time could not be measured by an eternal clock simultaneously ticking with the same beat throughout the universe. Our everyday notion that time was passing everywhere at the same rate, irrespective of the motion of the observer, would have to be abandoned.

Einstein's Bicycle

The implications of Einstein's new system of mechanics would prove to be dramatic. Objects that travel at close to the speed of light do not behave as we might expect. We will illustrate this in the guise of a thought experiment.

Imagine Albert Einstein cycling past us at a steady rate of half the speed of light (Figure 5.8). He shines his torch in front of him. His bicycle is equipped with a device that will measure the speed

Figure 5.8 Einstein riding a bicycle in Santa Barbara, California in 1933.

of the light that is being emitted by his torch. This could be some sort of interferometer, but we will just refer to it as his light meter. Einstein measures the speed of light without falling off the bike and, as usual, the rays in the torch beam are receding at the speed of light.

In our back garden, we also have a light meter that can measure the speed of light and we decide to measure the speed with which Einstein's torch beam is moving relative to us. We might expect that the torch beam would be moving at a speed that is equal to the sum of the bicycle speed plus the speed of light. In other words, the light would appear to move at one and a half times its normal speed. Our common-sense expectations have been developed through our experience of living in a world in which the laws of Newtonian mechanics are obeyed to a very good approximation. However, we only have direct experience of objects that move with very small relative velocities compared to the speed of light. It turns out that the obvious addition law of velocities that we take for granted is only approximately true. This approximation is an incredibly good one when we are considering objects moving at the low speeds that we are familiar with, but it becomes a very poor approximation as objects approach the speed of light. When we use our light meter to measure the speed of the beam emitted by Einstein's torch, we will find that it is moving at the speed of light. We know that this is the case because this thought experiment is really the Michelson–Morley experiment in disguise.

As Einstein explained, there is only one way to reconcile these two measurements of the speed of the torch beam. If we look closely at Einstein's light meter, we will see that the clock attached to it appears to be running more slowly than the clock attached to our light meter. If we look at the ruler attached to his light meter, it will appear to have shrunk by comparison to the ruler attached to our light meter. It would seem that the laws of Nature conspire to ensure that any measurement of the speed of light in a vacuum will always produce the same result, irrespective of the motion of the experimenter. If this were not so, then it would be possible to determine our velocity relative to the background of space, simply by measuring the speed of light, as Michelson and Morley attempted to do.

Incidentally, as Einstein races by, he will see us moving past him at half the speed of light. If he takes a look at our light meter, it will appear to him as though our clock is running slowly and as though our ruler has shrunk. This must be the case as the motion is relative and we cannot determine that one of us is moving and not the other, as encapsulated in Einstein's first postulate—the relativity principle.

Time Dilation

The apparent slowing down of the passage of time for a body in motion relative to us, such as Einstein on his bike, is called *time dilation*, and it is built into the laws of physics. It applies to time as measured in any conceivable way: grandfather clocks, digital watches, atomic clocks, egg timers, or the rate of stubble growth—it makes no difference.

You might think that this is all academic and it cannot have any real applications—nobody can cycle as fast as Einstein. But, in fact, it is an everyday occurrence. The first direct measurements of time dilation occurred in 1940. The electron has a heavy relative known as a muon. In many ways, the muon is very similar to an electron, but it is around 207 times as massive. It is also an unstable particle. Typically, a muon survives just 2 microseconds before decaying into other particles. Muons are created when cosmic rays hit atoms in the upper atmosphere and they travel at close to the speed of light when formed in these high-energy collisions. Even at such speeds, were it not for time dilation, the muons would not reach the ground before decaying. But, as Bruno Rossi (1905–1993) and David Hall demonstrated in 1940, muons do reach the ground in just the proportions expected due to time dilation. Their lifetimes are extended exactly as predicted by Einstein's theory.[7] Special relativity is now an integral part of fundamental physics and is routinely incorporated into our understanding of particle behaviour at laboratories such as the Large Hadron Collider (LHC).

At the LHC and other particle physics laboratories around the world, particles are routinely accelerated to within a whisker of the speed of light before being smashed into other particles. The resulting particle debris is then analysed in great detail. The newly created particles race away from the impact point at close to the speed of light and the sophisticated electronics and software within the particle detector identify each one and determine its energy and direction of flight. Most are highly unstable and short-lived—every species of particle survives for a characteristic and well-established length of time when at rest. Due to time dilation, the lifetimes of the high-speed particles are extended but always in accordance with special relativity. There are close to a billion collisions a second within the LHC and each collision is a test of special relativity. Einstein's theory passes these tests every time, and no-one expects it to fail in the foreseeable future.

Crossing the Light Barrier

It is one thing to declare that the speed of light is a universal speed limit, but what happens if we just keep on pushing an object faster and faster? Whether it be a spaceship or a proton in the LHC, can't we just keep on accelerating until we reach a speed that exceeds the speed of light? The answer turns out to be a definite *no*. Einstein realized that as an object approaches the speed of light, it becomes ever more difficult to raise its speed further. Effectively, the mass of the object increases as it approaches the speed of light so, when a force is applied, the change in speed becomes much smaller because of the corresponding increase in inertia. To accelerate the object all the way to the speed of light would mean that the mass of the object was pushed all the way to infinity, and this would require the input of an infinite amount of energy—which is, of course, impossible.

The consequences of this are dramatic, as Einstein explained. In another paper in 1905, Einstein used special relativity to work out a formula for how the inertia of an object increases as its speed increases. The formula emerged logically from the two postulates upon which he had built special relativity, so there was no doubting that it was correct if special relativity was correct. But the meaning of the formula was not immediately clear. Einstein's interpretation was typically brilliant, and it would introduce one of the most important unifying ideas in the history of physics. An approximate version of the formula looks something like this:[8]

$$\text{Inertial mass} = m + \frac{1}{2} m \left(\frac{v}{c}\right)^2 + \dots,$$

where v is velocity and c is the speed of light, so $\frac{v}{c}$ is the velocity as a fraction of the speed of light. The dots represent a sequence of other smaller terms, each multiplied by a power of $\frac{v}{c}$.

On the left of the equation is the inertial mass; this is the quantity that in Newtonian physics determines the relationship between force and acceleration via Newton's Second Law—the bigger the inertial mass of an object, the less acceleration it will receive when a force is applied to it. The first term on the right denoted by m is what we normally consider as the mass of the object and, in Newtonian physics, this corresponds exactly to the inertial mass. But, following Einstein, physicists refer to this as the *rest mass* of an object. All the other terms on the right involve velocity. At low speeds, the second term is far and away the biggest of these; it is the kinetic energy, the energy of motion in Newtonian physics, divided by the speed of light squared. According to Einstein's new relativistic analysis, there is a whole sequence of additional terms on the right, although the higher terms (represented by the dots) are small when the velocity is a small fraction of the speed of light. These are additional relativistic contributions to the kinetic energy.

Einstein argued that the best way to understand the formula is to postulate the complete equivalence of inertial mass with the total energy E, divided by c^2, so (after multiplying through by c^2) we can rewrite the formula as:

$$\text{Total energy} = E = mc^2 + \frac{1}{2} mv^2 + \ldots$$

Once this identification is accepted, the picture becomes much clearer. As we accelerate a body, we are increasing its energy and, because of the equivalence of energy and inertial mass, this increases the inertia of the body, which makes it more difficult to increase its velocity further. Therefore, applying a force to a rapidly moving body will increase its velocity by a smaller amount than applying the same force to the body when moving at a lower speed.

The punch line comes if we consider what the formula means when applied to a stationary body. In this case, all the kinetic energy terms on the right-hand side disappear. This makes sense because the body is not moving and, if $v = 0$, then all these terms are multiplied by zero. So we are left with the statement that for a stationary body the total energy is equal to the rest mass of the body multiplied by the square of the speed of light:

$$E = mc^2.$$

In Newtonian physics, a stationary object appears not to be carrying any energy at all, but Einstein interpreted his equation to mean that even a stationary body contains a quantity of energy equal to its rest mass multiplied by the speed of light squared. If Einstein was correct, then the amount of energy locked in the mass of material objects would be huge. And he most definitely was correct. This fact has had devastating consequences, but it is central to modern physics.

If mass is just another form of energy then the law of conservation of energy needs to be modified, since any change in mass must be accounted for and included in the balance. The amount of energy released in chemical reactions is very small compared to the amount of energy locked in the mass of the reacting atoms and molecules. It is so small that the consequent changes in mass had never been noticed before Einstein's startling revelation. Nuclear and particle physics were in their infancy, but it was already clear that nuclear interactions involve much larger amounts of energy. Physicists soon realised that when investigating these high energy areas of physics the equivalence of mass and energy must be taken into consideration.

Particle physicists talk in terms of the energy of the particles that are accelerated around their machines, rather than their speed. The energy of a particle is usually quoted in the convenient unit of the electron Volt. One electron Volt is the energy gained by an electron or other charged particle when it moves through a one Volt circuit. The equivalence of mass and energy means that this is also the best unit in which to specify the rest mass of a particle. For instance, the rest mass of a proton is just under 1 GeV, where GeV means 1 billion electron Volts.

When a proton is accelerated around the LHC, its speed barely changes but its energy steadily increases. A proton with an energy of 1 TeV (1000 GeV) in the LHC will be travelling around the machine within a whisker of the speed of light. To be precise, its speed will be 99.999% of the speed of light. After a few more minutes whirling around the LHC, the energy of the proton will have doubled to 2 TeV, but its speed will have barely increased at all. It will now be 99.9999% of the speed of light. The forces that are applied to the proton increase its energy, but its speed hardly changes. This is a perfect demonstration of the dramatic increase in the inertia of material bodies as their speed approaches the speed of light.

As soon as we accept that there is a maximum speed of interaction, then Einstein's theory of special relativity follows with a relentless logic. Some consequences of the speed of light barrier take a bit of getting used to. Nevertheless, the cosmic speed limit leads to theories of physics that are more consistent and philosophically appealing. Newton faced the criticism that his theories seem to imply instantaneous action-at-a-distance which is philosophically implausible. How can an object, such as the Sun, act on another object, such as the Earth, that is a great distance away without any time lag? This is what Newtonian physics appears to suggest. Newton's response was that his theories describe how the universe works, but they do not explain why it works in this way. Since Einstein, we believe that the interactions between two bodies are limited by the speed of light. This allows us to interpret how forces operate in much more reasonable and rational ways, as we will see.

The Fourth Dimension

Filby became pensive. 'Clearly,' the Time Traveller proceeded, 'any real body must have extension in four directions: it must have Length, Breadth, Thickness, and Duration. But through a natural infirmity of the flesh, which I will explain to you in a moment, we incline to overlook this fact. There are really four dimensions, three which we call the three planes of Space, and a fourth, Time. There is, however, a tendency to draw an unreal distinction between the former three dimensions and the latter, because it happens that our consciousness moves intermittently in one direction along the latter from the beginning to the end of our lives.'

HG WELLS, *The Time Machine* (1898)[9]

We have direct experience of living in three-dimensional space. It is intuitively obvious that there are three perpendicular directions to the space that our bodies occupy, and we could construct a coordinate system with three perpendicular axes. In addition, we have a sense that we are passing through time. We can imagine that time forms a continuum and as our clocks tick, the universe passes forwards in time along a fourth temporal dimension.

Imagine a fly in a cubic room. We could set up a coordinate system in which the x axis increases along one wall, the y axis along a second perpendicular wall, and the z axis runs from the floor to the ceiling. At each moment in time we could specify the position of the fly in terms of four coordinates. The x, y, and z coordinates determine the position of the fly in the three-dimensional space and the fourth coordinate is the time at which it is located at this position. The four coordinates will change in a continuous way as the fly buzzes around the room.

In this way, the four dimensions of space and time form the stage on which the drama of the universe is acted out. This intuition rests at the core of Newtonian mechanics. Newton did not describe physics in terms of four dimensions, but the idea that time could be considered as an extra dimension pre-dates relativity. Treating time as a fourth dimension is not, in itself, a product of Einstein's revolution; although famously part of relativity, it does not feature in Einstein's earliest formulations of the theory. The idea of uniting space and time as they are found in special relativity was first proposed in 1907 by the mathematician Hermann Minkowski (1864–1909), who had been one of Einstein's teachers.

> *The views of space and time which I wish to lay before you have sprung from the soil of experimental physics, and therein lies their strength. They are radical. Henceforth space by itself, and time by itself, are doomed to fade away into mere shadows, and only a kind of union of the two will preserve an independent reality.*[10]

Einstein initially dismissed the idea, possibly because he felt that it added unnecessary mathematical baggage, but he soon saw its tremendous potential for advancing the theory. The relationship between space and time is much more intimate in special relativity than in Newtonian physics. As we will see, the concept of a unified spacetime becomes very powerful. Unfortunately, Minkowski who died in 1909, did not live long enough to see the full fruits of his idea.

Another New Theory

Einstein realized that the cosmic speed limit would demand an even bigger upheaval in physics—a new theory of gravity. Newton's theory of gravity had been a cornerstone of physics for over 200 years and it had scored success after success, but there is a feature of Newton's theory that was incompatible with special relativity. The theory contains no mention of the time taken for the gravitational interaction to occur. Effectively, it assumes that the gravitational force is transmitted instantaneously. This aspect of Newton's theory had been questioned when it was first published, and Newton himself was well aware that it was philosophically untenable, but the theory worked so well that the need for a better theory had not arisen. Following the advent of special relativity built on the principle that information is never transmitted faster than the speed of light, the need for a new theory of gravity became obvious. This was simply a question of consistency.

The route to a new and improved theory of gravity was far from clear. One possibility might be to start with Newton's theory, which had been so successful for so long, and to modify it in line with special relativity to take account of the various effects that might arise as objects approach the speed of light. In addition to the Newtonian inverse square law term, there might be new terms to account for each relativistic effect. These might include: a term that would represent the delay in the application of the gravitational force due to its transmission being limited by the speed of light; a term that would represent the fact that mass and energy are equivalent and that the changing velocity of a body would change its energy and therefore also its mass; a term that would represent the energy in the gravitational field; a gravi-magnetic term that would be the gravitational equivalent of magnetism in electromagnetism; and so on.

In many branches of physics, the inclusion of additional terms to account for specific effects is the best approach to the construction of a model.[11] This would not be Einstein's approach to gravity. Einstein believed in working from fundamental physical principles, and this would mean discarding the entire apparatus of Newtonian gravity and starting from scratch. The end result would be a theory that looks like nothing ever previously constructed in the history of physics. It is almost as though humanity had been given the scientific theory of an alien civilization, or the physics of the twenty-first century was conceived a century ahead of its time.

A Happy Thought!

In 1907, Einstein had an idea that he would later describe as the happiest thought of his life. Einstein recalled that 'for a few days I was beside myself with joyous excitement'.[12] He explained the idea as a thought experiment. He imagined being in a lift whose cable had snapped; as the lift fell in the Earth's gravitational field, the occupants of the lift would feel weightless, just as though the Earth's gravitational field did not exist. In short, they would not feel the force of gravity. The reason that they would feel nothing is that the lift and everything in it, including every part of the body of each occupant, would fall with the same acceleration.

In a sense, there was nothing new about this idea; it is a feature of Newton's theory. According to Newton's Second Law of Motion, the acceleration produced by a force is equal to the size of the force divided by the mass of the object being accelerated. What this means is that a massive object resists being accelerated. We know this property as inertia. The bigger its mass, the greater the force required to change its motion. We have an intuitive understanding of this—it is much easier to throw a small stone than it is to hurl a large boulder. Now, in most cases where a force might be applied to an object, the size of the force will be completely independent of the mass of the object. For instance, the strength of the electromagnetic force on an object is proportional to the electric charge that the body carries (a very massive object might not feel any electromagnetic force if it is not electrically charged). But gravity is different. According to Newton's theory, the gravitational force that an object feels is proportional to its mass. In other words, a more massive object weighs more than a less massive object. Overall, the result is that, although a more massive object feels a greater gravitational force, it will also have a greater inertia, and this means that it will undergo the same acceleration in a gravitational field as a less massive object. The increased gravitational force due to the larger mass will exactly cancel with the increased inertia due to the larger mass. So what Newton's theory predicts is that all massive objects undergo the same acceleration in a gravitational field, with the result that Einstein's imaginary occupants in the plummeting lift feel weightless.

In Newton's theory, this feature of gravity seems almost accidental. What Einstein realized was that this unique attribute of gravitation could be used as a fundamental building block for a new theory. Einstein called it the *Equivalence Principle*—the equivalence of gravitational and inertial mass. Einstein could see that if the Equivalence Principle was used as the basis for a new theory, the notion of a force of gravity could be dispensed with altogether. Gravity would become a force without a force.

In the absence of any forces, all objects travel in the same way; they continue in a straight line at a constant speed. In a gravitational field, but in the absence of other forces, all objects will still travel in the same way because gravity affects all objects in the same way. Einstein realized that it would be more economical to discard the idea of a force of gravity acting on all objects in the same way and take the view that a massive object distorts the shape of spacetime and that other bodies then follow the straightest paths through this warped spacetime. Without gravity, spacetime is flat and objects travel along the shortest paths, which are straight lines; with gravity, spacetime is curved, and objects still travel along the straightest paths through this curved spacetime.

To make use of this idea, it would be necessary to encapsulate it mathematically. Prior to this time, Einstein had always assumed that it should be possible to describe fundamental physics using fairly elementary mathematics. It was the ideas underpinning the physics that mattered. Einstein had a single-minded determination to understand the physical universe and was not interested in mathematics for its own sake. Indeed, his teacher Minkowski is reputed to have referred to him as *a lazy dog* because of his attitude to mathematics. Clearly, this judgement was more than a little harsh. Einstein's neglect for his mathematical studies was actually a reflection of his general rebellious nature and the focus of his attention on physical problems. Now, with the need to develop a new theory of gravity, Einstein knew that he would require some sophisticated mathematics. In Chapter 6, we will see how Einstein moulded a new universe out of the geometry of curved space.

6

Let's Do the Time Warp Again!

Of meridians and parallels
Man has weav'd out a net, and this net throwne
Upon the Heavens, and now they are his owne.
  JOHN DONNE, *An Anatomie of the World First Anniversary* (1611)

Weaving the Net

Einstein needed to find a way to navigate his way around curved spacetime. This was a similar problem to that faced by the cartographers of earlier centuries. There is a fundamental difficulty in producing an accurate map of the world: the Earth is a globe, so it has a curved surface, whereas a map is a flat piece of paper. If the surface of the Earth were a cylinder there would be no problem because a cylindrical map of its surface could be unrolled to lie flat. But there is no consistent way to take the features of a spherical surface and place them onto a flat map without some distortion. Once the necessity of some distortion is accepted, however, there are many ways to create a map, and each has its own characteristics. The particular construction chosen by a cartographer depends on the purpose of the map and which of its features are deemed most important. This may be as much a matter of politics as science.

Cartographers use techniques that resemble those that artists use to produce realistic perspective in their drawings. The artists' canvas may be thought of like a window looking out on the scene that is to be represented.[1] The drawing is then constructed by imagining rays of light travelling in straight lines from each point in the scene to the artist's eye or the viewpoint of the drawing. The drawing will appear realistic if it looks like a cross-section of these projected light rays, so the challenge for the artist is to reproduce this cross-section. Albrecht Dürer (1471–1528) offered a concrete illustration of this procedure in a famous woodcut shown in Figure 6.1. In the woodcut, a taut string represents the light rays, and is secured to the wall at the viewpoint of the drawing that is under construction. The other end of the string is attached to a point on a lute that is being drawn. The position at which the string crosses the frame of the canvas is measured and then this point is transferred to the canvas. Point by point, the image of the lute is built up as a cross-section of the light rays projected from the lute.

Just as projection can be used to produce an image of a solid object, such as a lute, on a two-dimensional canvas, so projection can be used to represent a spherical globe on a map. Cartographers have devised various methods to achieve this. One

Figure 6.1 Albrecht Dürer illustrating the art of perspective drawing in his *Instructions in the Art of Measurement* (1525) by constructing the image of a lute point by point.

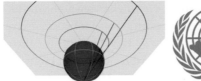

Figure 6.2 Left: Stereographic projection of points on a sphere to a planar map. Moving outwards from the North Pole the three circles are the Tropic of Cancer, the Equator, and the Tropic of Capricorn. Right: The logo of the United Nations is based on a stereographic projection of the globe.

is stereographic projection, as illustrated in Figure 6.2, Left. Imaginary lines are drawn from the South Pole through the globe to the map, and the point where each line crosses the globe's surface is plotted at the point where the line reaches the map. In other words, the map is a cross-section of the lines projected from the South Pole. Lines of latitude, such as the Equator, project to concentric circles on the map, and all points in the northern hemisphere are projected inside the circle representing the Equator. Moving radially outwards from the centre, equally spaced latitude circles are separated by increasing distances on the map. By contrast, lines of longitude are projected to straight lines radiating out from the North Pole. If the globe were a transparent sphere painted with opaque continents and lines of latitude and longitude, the map would be the silhouette produced by a light bulb positioned at the South Pole.

Maps of the Earth in stereographic projection are unusual. One familiar example is seen on the flag of the United Nations (Figure 6.2, Right), and it is a good illustration of the properties of this projection. The Arctic is in the centre of the map, and it appears without much distortion, but the southern hemisphere is stretched outwards. As we move radially from the centre of the map, equal distances on the globe correspond to increasing distances on the map, so the southern land masses look huge compared to the northern continents. Antarctica is completely missing from the map as it would stretch right around its perimeter. Although a map such as this might serve its political or decorative function well, it would not be much use for navigating the seven seas. There are other projections that have proved much more suitable for these purposes.

The most familiar world maps use a projection that was invented by the Flemish cartographer Gerardus Mercator (1512–1594). Mercator used a type of cylindrical projection that can be imagined as follows. A cylinder of paper is wrapped around the globe, then lines are projected from the centre of the globe through each point on the surface to the location on the cylinder where the point is to be plotted (Figure 6.3, Left). The paper cylinder is then unrolled to produce a flat map (Figure 6.3, Right).

This central cylindrical projection, as it is called, produces extreme distortions in distance as we move towards the Poles. Mercator compensated for this by scaling distances in the north–south direction. Although this reduces the distortion, some inevitably remains. Maps using the Mercator projection are accurate near the Equator, but regions far from the Equator

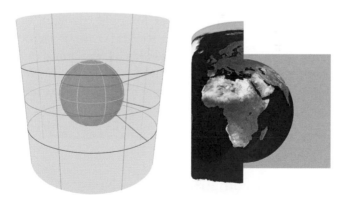

Figure 6.3 Left: The Mercator projection is produced by projecting from the Earth's centre through the Earth's surface and onto a cylinder wrapped around the globe. Right: The map is then created by unrolling the cylinder from around the globe.

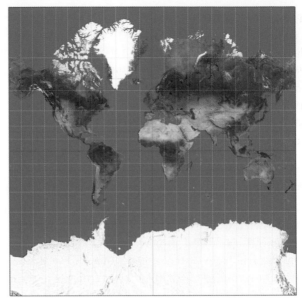

Figure 6.4 In the Mercator projection, Greenland appears larger than South America even though in reality it has only one-ninth of the area.

appear abnormally enlarged, so Greenland looks about the size of South America even though it has only one-ninth of the area. A Mercator projection of the world is shown in Figure 6.4. Mercator introduced this projection for his world map of 1569, and it soon became the standard way of representing the world.

Despite the success of the cartographers, mathematicians saw the world rather differently. Their approach to geometry was much more abstract and it was not until the nineteenth century that there was a convergence of ideas.

Geometries Old and New

In the third century BC, the Greek mathematician Euclid wrote *The Elements* in which he proved the most important results of geometry. These *theorems*, as they are called, were built up step by step from a small set of *axioms* that were taken as irrefutable truths. Euclid began with simple results, such as a proof that the angles of a triangle sum to two right angles or 180°, and ultimately reached much more complicated results about the structure of the regular solids. For thousands of years the notion of geometry meant the geometry of Euclid; there was a universal belief in the existence of a direct correspondence between Euclid's geometry and the structure of the real world. It was considered so obvious that the results of Euclid must hold in all circumstances that they were never questioned.

Puzzle 8 A sphere can be divided into eight equal parts by three circles; if one is the Equator, then the other two circles cross the Poles in perpendicular directions, as shown in Figure 6.5. Each of the eight equal parts of the sphere is a spherical triangle. What is the sum of the angles of one of these triangles?

Figure 6.5 A sphere that has been divided into octants by three great circles.

> **Answer to Puzzle 8** Each of the three angles of the spherical triangle is a right angle. The sum of the angles is therefore $270°$.

Yet the geometry of Euclid is not valid on a sphere. As Puzzle 8 illustrates, on a sphere the angles of a triangle do not add up to $180°$. Indeed, the geometry of a sphere is quite different to the geometry of Euclid. This is because one of Euclid's axioms—the parallel postulate—does not hold on the sphere. According to the parallel postulate, if we take a straight line and a point that is not on the line, then we can always draw a unique line through the chosen point that is parallel to our line. This sounds quite reasonable, or at least it did for 2000 years. In fact, this axiom holds on a flat surface, but not on a curved surface such as a sphere. If we take two lines of longitude, they might look parallel at the equator but they meet at both the North and South Poles (Figure 6.6). There are no parallel lines on a sphere.

Because this axiom does not hold for the geometry on a sphere, Euclid's results that rely on this axiom will not hold on a sphere. This is the case with his proof that the angles of a triangle sum to $180°$ (Figure 6.7). Euclid's geometry is actually a very special geometry; it is the geometry of flat space.

A Strange New Universe

It is perhaps surprising, but mathematicians did not discover the existence of *non-Euclidean* geometries until the early years of the nineteenth century and, even then, it was not through the contemplation of spherical geometry. Geographers had been producing accurate maps of the globe for several centuries, and astronomers had been mapping the heavens since antiquity, yet no-one had realized that the geometry of the sphere could be constructed in a manner that was comparable but distinct from the Euclidean geometry that describes flat space.

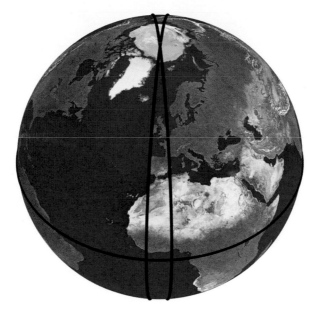

Figure 6.6 Two lines of longitude. They appear parallel at the equator, but meet at the North and South Poles.

There were mathematicians who had refused to give the parallel postulate the same status as the other axioms of geometry, and several had devoted their lives to fruitless attempts to prove that it was not truly independent, by showing that it could be derived from the other simpler axioms. One such mathematician was the Hungarian Farkas Bolyai who became seriously alarmed when his son János also began to take an interest in the problem. Farkas warned him:

> *For God's sake, please give it up. Fear it no less than the sensual passion, because it, too, may take up all your time and deprive you of your health, peace of mind and happiness in life.*[2]

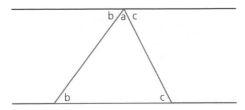

Figure 6.7 To prove that the angles of a triangle sum to 180° we can proceed as follows. Extend the base of the triangle with a straight line that extends to infinity. Draw a line parallel to this line through the third vertex of the triangle. Then the two angles labelled b in the diagram are equal (they are alternate angles). The two angles labelled c are also equal. The sum of angles a, b, and c is 180° because they form a straight line. But these are also the three angles of the triangle. Therefore the angles of the triangle must sum to 180°. The proof clearly relies on the parallel postulate.

János did not heed his father's advice, however, and in 1823 he wrote to his father:

> *I have discovered such wonderful things that I was amazed . . . out of nothing I have created a strange new universe.*

János Bolyai (1802–1860) had made a breakthrough that would rock the world of mathematics. János had realized that Euclid's parallel postulate is independent of his other axioms, so it cannot be proved from any simpler statements. But he also found that it is optional. If we accept it, as Euclid did, then all the results of Euclidean geometry follow. If we do not accept the parallel postulate, then our axioms and the results that we derive from them correspond to other geometries. Whereas Euclid's geometry describes flat space, these other geometries describe space that need not be flat.

János was not alone. A young Russian called Nikolai Lobachevsky (1792–1856) had independently made the same momentous discovery. When the world's leading mathematician Carl Friedrich Gauss (1777–1855) was informed about the work of Bolyai and Lobachevsky, he revealed that he had also arrived at the same conclusions but had chosen to keep quiet.[3] These three mathematicians had seen the problem differently from all their predecessors, and each had constructed an alternative non-Euclidean geometry known as *hyperbolic* geometry.

Whereas Euclidean geometry is the geometry of flat space, hyperbolic geometry is the geometry of a saddle or a curved funnel known as a *catenoid* (Figure 6.8). As we have seen, in Euclidean geometry the angles of a triangle always sum to 180°, while in spherical geometry the angles of a triangle always sum to more than 180°. So what about triangles in hyperbolic geometry?

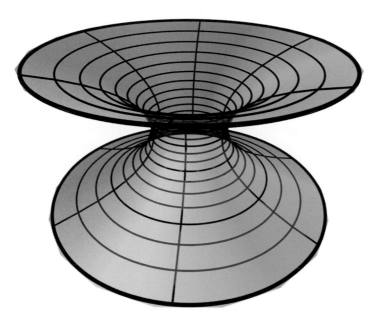

Figure 6.8 Hyperbolic geometry is the geometry of a saddle or curved funnel known as a catenoid. (The catenoid extends outwards to infinity, so in the image it has been truncated at the top and the bottom in order to fit on the page.)

Gravity

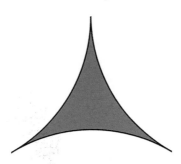

Figure 6.9 An example of a triangle in hyperbolic geometry. Each of the angles of this particular triangle is less than 60°, so the sum of the angles is less than 180°.

Figure 6.9 shows an example of a triangle that might be drawn on a hyperbolic surface such as a catenoid; each of its angles is less than 60°, so their sum is less than 180°. In hyperbolic geometry, the angles of a triangle always sum to less than 180°.

Curvaceous Figures

We have an intuitive understanding of the difference between a flat surface and a curved surface, and it is clear that the difference between the three geometries—flat, spherical, and hyperbolic—has something to do with curvature, but it took Gauss to make this distinction precise. Consider a hyperbolic surface. Around each point, the surface curves in two perpendicular planes and the centres of curvature are in opposite directions, as shown for one particular point in Figure 6.10. The arrows in the figure point towards the centres of curvature and, as they indicate, the centres of curvature are in opposite directions. This corresponds to negative curvature,[4] and it is true throughout a hyperbolic surface; it is the defining feature of hyperbolic geometry.

Now consider the more familiar example of a spherical surface. In Figure 6.11, two perpendicular circles are drawn through a representative point on the surface. In this case, the centres of

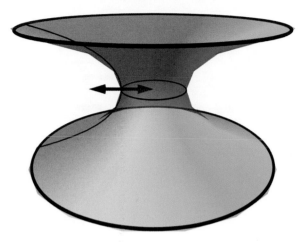

Figure 6.10 The arrows indicate the directions towards the centre of curvature at a point on the hyperbolic surface.

curvature are both in the *same* direction, which is towards the centre of the sphere. So, whereas a hyperbolic surface is negatively curved, the sphere has positive curvature. Gauss showed that the hyperbolic, flat, and spherical geometries correspond to the three possible geometries of two-dimensional surfaces with constant curvature.

What Is the Shape of Space?

The existence of these new geometries raised the question of the true geometry of space; Euclidean geometry could no longer be accepted without question. In the 1820s, Gauss was commissioned to survey the German Kingdom of Hanover, a project that he completed by triangulating various prominent locations. Gauss was analysing the mathematical properties of curved surfaces during the same period, and in 1827, he published his most important work on the subject, *General Investigations of Curved Surfaces*.[5] This has led to the suggestion that Gauss attempted to

Figure 6.11 The arrows indicate the directions towards the centre of curvature at a particular point on the sphere; in this case the two arrows point in the same direction—inwards.

determine the geometry of space in the Earth's vicinity while conducting his survey. According to the story, Gauss measured the angles of a large triangle formed by three of Hanover's most prominent peaks: the Brocken, Hohenhagen, and Inselberg mountains,[6] and when the angles summed to 180°, Gauss concluded that the space surrounding the Earth is flat. Unfortunately, there does not seem to be any documentary evidence to show that Gauss attempted to answer this question, and he would almost certainly have realized that his instruments were insufficiently sensitive to detect any deviation from Euclidean geometry.

The mathematical exploration of curved surfaces was transformed by one of Gauss's pupils. Bernhard Riemann (1826–1866) was the son of a poor Lutheran pastor. In 1846, his father scraped together enough money to send him to the University of Göttingen, where Riemann intended to acquire a degree in theology. After attending Gauss's lectures, however, Riemann fell in love with mathematics and, with his father's blessing, changed the direction of his studies. Riemann, who was penniless, painfully shy, and home-sick, devoted himself entirely to mathematics. Mentored by Gauss, he rapidly developed into one of the truly great mathematicians of the nineteenth century. Most mathematicians tread carefully and take a rather cautious approach to their subject, but Riemann was different. He had the vision to open up sweeping new vistas of mathematics. In 1853, Gauss asked Riemann to prepare a thesis for his habilitation examination—a high-level qualification required by lecturers at German universities—and it was delivered the following year with the title: *On the Hypotheses which Underlie Geometry*. Riemann realized that hyperbolic geometry was just the start of a revolution, and in this short paper he set out principles that would transform geometry, extending it into hitherto unimagined realms. Riemann presented methods that would enable mathematicians to describe surfaces that curve in arbitrary ways and set out a programme for the development of a generalized geometry of curved spaces in any number of dimensions.

Riemann made many other wide-ranging contributions to mathematics, even though his personal life was blighted by depression and the loss of family members that were so dear to him. The ravages of long-term poverty did nothing for his health, and Riemann finally succumbed to tuberculosis in 1866, at the age of just thirty-nine, while visiting the village of Selasca on Lake Maggiore, in Italy.

Riemann's key to a general understanding of arbitrary surfaces is closely related to cartography. There are many ways in which to create a map of the world, as we have seen. Each projects the features of the globe onto a flat surface, and each projection produces its own characteristic distortions of these features.

This distortion means that the scale varies from region to region across the map, such that a line between two points in one region might represent a different distance on the globe to a line of the same length in another region of the map. Any navigator using the map requires a set of instructions that explain how the scaling varies, so these distortions can be compensated for. In effect, as the navigators move their rulers over the map, they must imagine that their length is constantly changing. In Mercator's maps, for instance, regions near the poles appear comparatively much larger than regions near the Equator. So in this projection a ruler is longer at the poles than it is near the equator. One hundred kilometres might be represented as 1 centimetre close to the equator, but it could be represented as 3 centimetres towards the poles.

Figure 6.12 The entire infinite hyperbolic plane can be mapped into a disc (although this was first proved by the Italian mathematician Eugenio Beltrami, the hyperbolic disc model is often known as the Poincaré disc). The illustration shows a tessellation of triangles in the hyperbolic plane. Each triangle has the same area, but the projection onto the disc distorts the size of the triangles, just as Mercator's projection distorts the area of the continents.

Puzzle 9 What is the sum of the angles of each of the triangles in the hyperbolic tessellation in the disc shown in the Figure 6.12?

HINT: It is a regular tessellation. Consider the puzzle about tessellations in Chapter 2.

Riemann adopted a similar approach to the analysis of arbitrary curved surfaces in any number of dimensions; he devised a method by which the surface could be projected onto flat space (with the same number of dimensions), along with a recipe for how the distance between any two points in the surface could be calculated. All the necessary information about distances is encapsulated in a mathematical object known as the metric, and its form is determined by the shape of the space that is being considered. For instance, the metric of flat space is different to the metric of a sphere, which is different to the metric of hyperbolic space. Riemann's ideas were developed by other mathematicians in the later years of the nineteenth century. This mathematical toolkit is now known as Riemannian geometry. It would prove to be the perfect box of tricks for Einstein's theory of gravity.

Answer to Puzzle 9 The angles that meet at each point must sum to $360°$. Eight triangles meet at each vertex in the tessellation (four black and four white). Each triangle in the tessellation is identical to the rest, therefore as $8 \times 45 = 360$, the eight angles that meet at each vertex must each be $45°$. The sum of the angles of each triangle is therefore $3 \times 45° = 135°$.

Back to Einstein

Galileo had demonstrated the counterintuitive fact that all objects fall with the same acceleration under gravity irrespective of their mass. In Newtonian physics, mass has a double role; it is responsible for our inertia and, thereby, diminishes our acceleration when a force is applied, but it also generates the force of gravity. The cancellation between these two properties of mass means that all objects are affected by gravity in exactly the same way, whatever their mass, and this accounts for Galileo's observation. Newton accepted that mass plays this dual role without comment.

Einstein saw it differently. He thought that this equivalence was so striking that it must be a fundamental principle of the universe, not just a lucky coincidence. As we have seen, Einstein realized that when falling under gravity (such as in a lift whose cable has snapped), if no other forces are acting on us then we do not feel any force at all. Every part of our body and everything else around us falls in exactly the same way. Einstein suggested that, in free fall, gravity is undetectable as all the laws of physics operate just as they would in empty space far from the gravitational effects of any massive bodies. For this reason, Einstein believed that it must be possible to describe gravity without using Newton's concept of a gravitational force.

Einstein would achieve this by the introduction of spacetime curvature into physics. Minkowski had shown the potential impact that geometry could have on physics by sewing space and time together into four-dimensional spacetime. In Einstein's description of mechanics, which is known as special relativity, spacetime is the flat background through which objects such as particles, planets, and spaceships move. Einstein would now incorporate gravity into the theory by transforming this static background into a dynamic part of the theory. His idea was that matter would shape spacetime, and then spacetime would determine how matter would move. In Newton's theory, the motion of a projectile is determined by the force of gravity. But as all free-falling projectiles follow the same path through spacetime, Einstein thought it economical to discard the gravitational force and view spacetime as curved by gravity. The free-falling projectiles would then simply follow straight paths through this curved spacetime.

A straight path through curved space sounds like a contradiction, but it is fairly easy to generalize our flat space notion of a straight line. In flat space, a straight line is the shortest path between two points. Similarly, we can determine the shortest path between two points on a curved surface. These paths are known as geodesics and, when space is curved, they are the equivalent of straight lines. The Mercator maps of the world that we are familiar with distort our perception of distance on the globe. Lines of longitude and latitude both appear straight on these maps. But whereas lines of longitude are the shortest paths between the points that they connect, lines of latitude (other than the Equator) are certainly not the shortest routes. So, although lines of longitude are geodesics, lines of latitude are not. Typically, the lines that appear straight on the map are not actually the shortest routes. For instance, the shortest route when flying from London to Tokyo is to follow a great circle route that takes an aircraft well into the Arctic, even though Tokyo is much further south than London, as can be seen by comparing Figures 6.13 and 6.14.

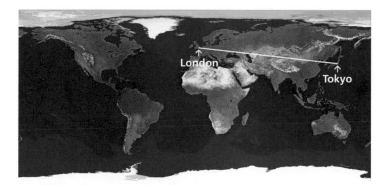

Figure 6.13 Map of the Earth produced using a Mercator projection. What appears to be a straight line between London and Tokyo is actually a much longer route than that shown in the figure.

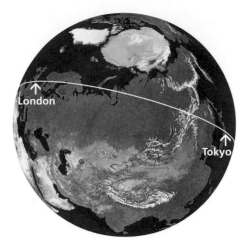

Figure 6.14 The shortest path between London and Tokyo follows a great circle route that goes via the Arctic. (A great circle is one that divides the Earth into two equal hemispheres.).

Einstein proposed that bodies falling freely in a gravitational field could be treated as though they were following a geodesic in curved spacetime. This would update the idea of Newton's natural inertial motion to the context of Einstein's new theory. According to Newton's First Law, if no forces are acting, objects move at a constant velocity in a straight line through space. In Einstein's theory, inertial motion also includes motion in a gravitational field. According to Einstein, if no forces are acting other than gravity, then objects move along a geodesic in spacetime.

This was just the first half of the story. It would explain how bodies travel through curved spacetime. But what Einstein needed was a relationship that would determine exactly how mass curves spacetime. Finding this relationship would involve Einstein in an arduous intellectual struggle.

Blowing Bubbles

Einstein's aim was somehow to use spacetime curvature to replace Newton's tried and tested concept of gravitational forces.

At first sight, this seems like an outlandish idea. How could curvature play an equivalent role in physics to a force? An everyday analogy shows that this is not really as crazy as it might seem. If two rings are placed together in soapy water until they are covered in a soapy film and then gradually separated, a bubble may form between the rings (Figure 6.15). The bubble is shaped like part of the hyperbolic surface known as a catenoid.

The bubble's shape is determined by the surface tension in the soapy water. At each point on the surface, the surface tension generates forces that act towards the centres of curvature. On a hyperbolic surface, the centres of curvature at each point are in opposite directions, as shown in Figure 6.10. This means that the surface tension forces can exactly balance at each point, as they must if the surface is to be stable, which is why the bubble forms this particular shape.

A second example is even more familiar. When we blow soap bubbles into the air, they form spheres. The direction towards the centre of curvature at each point on the surface is now towards the centre of the sphere. This means that the surface tension is producing an inward force at each point on the spherical bubble (Figure 6.16). The inward force squeezes the air within, raising its pressure until the air inside the bubble exerts a force that is sufficient to resist any further contraction of the bubble. This is why bubbles sink; the air within the bubble is slightly denser than

Figure 6.15 A soap bubble stretched between two rings. The bubble is a hyperbolic surface.

Figure 6.16 Spherical bubble.

the air outside. The difference in the air pressure between the inside and the outside of the bubble balances the inward pressure due to the surface tension of the bubble. (When producing the hyperbolic bubble, the rings were open at each end of the surface. This allows air to circulate freely, so the air pressures on the inside and outside of the surface are the same.)

So, as the bubble analogy shows, Einstein's intention of relating curvature to forces was not unreasonable. But to formulate his new theory, Einstein needed a better understanding of curved space and a thorough grounding in the mathematical machinery developed by Riemann and the later mathematicians who built on his work. This was an arcane topic that was unfamiliar to physicists at the beginning of the twentieth century. Fortunately, one of Einstein's close friends, Marcel Grossmann (1878–1936), was a mathematician who was an expert in the subject. Einstein spent several years in Zurich working with his friend to master the intricacies of Riemannian geometry. Einstein's goal was a recipe for calculating the metric of curved spacetime; what he needed was an equation that would encapsulate how a massive object warps the spacetime region that surrounds it.

The Messenger of the Gods Dances a Jig

Einstein announced to the world the final form of general relativity in November 1915. He had found the precise relationship he was seeking. On one side of the equation is a mathematical object that describes the distribution of mass and energy in a region of spacetime.[7] On the other side is a mathematical object that encodes the curvature that this mass and energy generates in this region of spacetime.[8] It is as simple as that. The equation is known quite naturally as Einstein's equation. Given a particular distribution of matter, it determines how spacetime will be warped. A solution of the equation determines the metric of spacetime. In other words, it describes how the length of a ruler and the pace of a clock will vary throughout spacetime. This, in turn, determines the trajectories followed by objects as they wend their way through spacetime.

The solar system had been used as a laboratory to test Newtonian gravitation for several hundred years, and Newton's theory had scored a long sequence of triumphs. If general relativity were to be a viable theory, then, as Einstein well knew, it would have to reproduce these great successes. Solving Einstein's equation is easier said than done, however. The equation is actually ten coupled equations that must all be simultaneously satisfied. Solving Einstein's equation for a general distribution of matter is still beyond the mathematical technology of today. By contrast, it is much easier to find approximate solutions, and this is perfectly adequate for testing the theory in situations where gravity is not too intense and for objects that are moving slowly compared to the speed of light, such as within the solar system where Newton's theory had scored its greatest successes. In making these approximations, it is natural to convert the esoteric picture of objects following geodesics through curved spacetime into the more familiar imagery of Newtonian mechanics.

Einstein could see when he undertook these calculations that to a first approximation, general relativity looked exactly the same as Newtonian gravity—which was good news. According to the theory, spherically symmetric objects that are not too dense act on each other with a force that diminishes as the inverse square of distance, just as Newtonian gravity says, which means that all the great successes of Newton's theory also automatically apply to general relativity. This was a great first step. Nonetheless, simply reproducing all the results of an already successful theory fell far short of Einstein's ambition. Einstein needed to examine the theory beyond the Newtonian approximation in search of results that differed from the Newtonian predictions, so that the two theories could compete head to head in their explanation of the physical universe.

When the analysis of general relativity goes one step further, there is a small additional term that is equivalent to a force that diminishes as the inverse fourth power of distance. This represents a clear distinction to Newtonian gravity where the force of gravity corresponds to an inverse square law and nothing else. Therefore, this new term derived from general relativity offers the possibility of testing the theory in what is known as the Post-Newtonian approximation.

As Newton proved, an exact inverse square law will result in elliptical orbits and, in this way, Newton's theory of gravity accounts for Kepler's First Law of Planetary Motion, a purely descriptive law that Kepler derived from Tycho's observations. The effect of a small additional force would be to cause the direction of the axis of the ellipse to change gradually. In other words the orbit would precess. This was clear without making any detailed calculations. But what Einstein wanted to know was how big this effect would be. How fast would the axis of the ellipse precess? In the solar system, the effect of this extra term is largest in the case of Mercury, because Mercury is closest to the Sun, which means that Mercury is deeper in the Sun's gravitational well. Mercury also travels faster than any of the other planets and this also produces greater relativistic effects. When Einstein performed this critical calculation, he found that the additional term due to general relativity would cause Mercury's orbit to precess by precisely 43 seconds of arc per century.

This was the moment when Einstein knew his theory was a phenomenal success. The size of the precession produced by these relativistic effects exactly matched the known discrepancy in Mercury's orbit—the discrepancy that Le Verrier had attempted to plug by hypothesizing the existence of the inner planet Vulcan. Now Einstein knew the real reason for the precession of Mercury's orbit. It was nothing to do with an invisible planet but simply the consequence of the curvature of spacetime predicted by his revolutionary new theory of gravity. In the words of his friend and biographer, Abraham Pais:

> *This discovery was, I believe, by far the strongest emotional experience in Einstein's scientific life, perhaps in all his life. Nature had spoken to him.*[9]

Heroism on the Eastern Front

Karl Schwarzschild (1873–1916) was a mathematician and astrophysicist with wide-ranging interests. When the First World War broke out in August 1914, Schwarzschild was forty years old and a professor at Germany's prestigious University of Göttingen. Despite his age and his unsuitability for military conflict, he immediately volunteered for military service. Initially, he was put in charge of a meteorological station in Belgium. He was then stationed with an artillery unit in France, calculating missile trajectories before being posted to the Eastern Front and the conflict with Russia. On the Eastern Front, Schwarzschild began to suffer from a rare and extremely painful autoimmune disease of the skin known as pemphigus.

Somehow, amidst the insanity of war, deafened by the roar of shells and with painful, blistering skin, Schwarzschild was able to focus his mind on Einstein's general relativity published in November 1915. Einstein's new theory was highly original, as we have seen, and employed ideas and mathematics that had never previously been used by physicists. Furthermore, the fundamental equations of the theory were very complicated and difficult to solve exactly. Einstein had only been able to find approximate solutions, which he had used to calculate the precession of the orbit of Mercury.

Schwarzschild wrestled with the equations in the isolation of the trenches and, despite the novelty of the mathematics, within a few weeks he had found the most important solutions of Einstein's equation. Schwarzschild had dramatically simplified the mathematical analysis by considering a very symmetrical situation, a spherically symmetrical mass. His solutions describe the shape of spacetime outside and inside a perfectly spherical body such as a star or a planet. On 22 December 1915, he wrote to Einstein to tell him of his discovery. Schwarzschild's concluding remarks demonstrate his incredible courage in the most desperate of circumstances: 'As you see, the war treated me kindly enough, in spite of the heavy gunfire, to allow me to get away from it all and take this walk in the land of your ideas.'[10]

Einstein was taken aback by the solution that Schwarzschild had found. He replied: 'I have read your paper with the utmost interest. I had not expected that one could formulate the exact solution of the problem in such a simple way.'[11]

Schwarzschild's skin condition worsened and in March 1916, he was removed from the front. Two months later, on 11 May, Schwarzschild died.

Total Eclipse of the Sun

On 11 November 1918, the guns finally fell silent on the Western Front. Following the armistice, the British astrophysicist Arthur Eddington (1882–1944) supported the idea of an expedition to test a prediction of Einstein's theory of general relativity. Eddington, who was a quaker, felt that the expedition would offer a great opportunity for reconciliation after the horror of the First World War. British astronomers would be testing a theory by a German physicist that could overthrow the theory of gravity devised by the greatest of all British scientists. According to general relativity, the path of a beam of light should bend when passing through the warped space close to a massive object such as the Sun. This meant that the position of a star would appear to shift very slightly when it was close to the edge of the Sun's disc as the Sun's gravity would bend the path taken by the star's light. Viewing the stars close to the Sun is almost impossible, however, as they are drowned out by the Sun's intense light. Fortunately, on Earth there are rare occasions when this is not a problem—during total eclipses of the Sun.

In the nineteenth century, solar eclipses had given astronomers the opportunity to confirm Newton's theory of gravity by searching for the elusive planet Vulcan. Now, in the twentieth century, astronomers would take advantage of a solar eclipse in the hope of overturning Newton and confirming the new theory of Einstein. Eddington travelled to the island of Príncipe, off the west coast of Africa, to photograph the region around the Sun during the total solar eclipse of 29 May 1919, and the photographs did indeed show a small shift in the positions of the stars close to the Sun. The light-bending power of gravity had been witnessed. Eddington heralded the results as a triumph for science and announced to the world's press the sensational confirmation of the theory of general relativity. From this moment, Einstein's life changed completely. He would be feted worldwide as the greatest intellect on the planet for the rest of his life.

In reality, the eclipse results were not very accurate and their agreement with general relativity was not as conclusive as Eddington may have suggested. Nonetheless, he was correct. We now have beautiful pictures that offer a dramatic illustration of this remarkable feature of gravity and its interpretation by Einstein.

Figure 6.17 The mass of the relatively nearby galaxy cluster MACSJ0138.0-2155, at the centre of the figure, forms a gravitational lens that produces the wispy orange warped and magnified images of a far more remote galaxy that lies behind it, 10 billion light years away.

Figure 6.17 shows a cluster of galaxies whose immense gravitational field bends the light rays from even more distant galaxies, so that they appear as distorted wispy arcs; their images are magnified, brightened, and distorted by the curved space around the cluster, and in some cases, there are multiple images of the same distant galaxy. The curved space around the galaxy cluster acts like a lens and is, indeed, referred to as a gravitational lens (Figure 6.18).

As an added bonus, gravitational lensing offers a great way to measure the amount of mass within the object that is responsible for bending the light. This has given astronomers a valuable tool to work out the mass of a cluster of galaxies. The results are rather surprising, as they indicate that there is far more matter in the universe than we can see. We will return to this issue in Chapter 9.

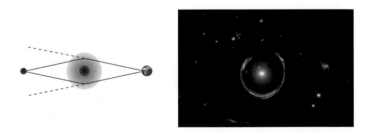

Figure 6.18 Left: Schematic diagram of a gravitational lens. The light from a distant object bends around a very massive intervening object such as a galaxy or a cluster of galaxies. From Earth, multiple images of the more distant object may be seen. If the alignment is perfect the object will appear as a ring around the intervening mass. Right: The image of a distant galaxy is almost warped into a ring around a galaxy at an intermediate distance. (LRG 3-757, © HST, NASA/ESA.)

Doing the Timewarp!

It's astounding
Time is fleeting
Madness takes its toll
But listen closely
Not for very much longer
I've got to keep control
I remember doing the Time Warp
Drinking those moments when
The blackness would hit me,
And the void would be calling
Let's do the Time Warp again
Let's do the Time Warp again

RICHARD O'BRIEN, *Time Warp* (1973) [12]

According to general relativity, mass not only curves space—it also warps time. One way to measure the passage of time is to use the regular beat of an electromagnetic wave, such as a gamma ray. In 1959, Robert Pound (1919–2010) and Glen Rebka (1931–2015) devised a classic experiment to test Einstein's prediction that time is warped in a gravitational field based on firing gamma rays down Harvard University's 20-metre Jefferson Tower.[13]

Pound and Rebka used gamma rays that are emitted with a very sharply defined frequency by the nuclei of a radioactive iron source. At the bottom of the tower they positioned a second thin piece of iron, beneath which was a gamma ray detector. Each gamma ray photon emitted by an iron nucleus at the top of the tower should have exactly the right energy to excite another iron nucleus at the bottom. Except that according to general relativity, falling in the Earth's gravitational field would change the frequency of the gamma radiation very slightly as the photons would undergo a blue shift on their journey to the bottom of the shaft. (Another way to look at this is that the photons gain energy as they fall in the gravitational field and, as the energy of a photon is proportional to its frequency, this increase in energy translates into an increase in frequency.) The predicted small shift in frequency meant that after falling down the tower the gamma rays would have the wrong energy to be absorbed by the iron nuclei. They would therefore travel straight through the iron to the detector below. But, if the theory were wrong, then no shift would occur. In this case, the gamma rays would be absorbed by the iron nuclei, and they would not be detected below.

The really clever feature of Pound and Rebka's experiment was that the iron absorber could be slowly driven upwards or downwards. When the velocity was just right this would produce a tiny Doppler shift that would exactly compensate for the gravitational shift in frequency, producing a peak in the radiation mopped up by the iron absorber and, therefore, a trough in the gamma ray photons reaching the detector below. This ingenious arrangement enabled Pound and Rebka to detect the tiny gravitational blue shift produced by the Earth's gravitational field and show that it matches the predictions of general relativity. In the Earth's relatively weak gravitational field, the effect is tiny, amounting to a shift in frequency of around one part in a thousand trillion (10^{-15}). But this shift in frequency shows that time passes more slowly close to the Earth's surface due to its gravity, just as Einstein predicted.

Where Are We?

Although Pound and Rebka's experiment confirmed the time-warping predictions of general relativity, it would be reassuring to test such counterintuitive effects in an everyday environment. So in 1971, Joseph Hafele (1933–2014) and Richard Keating (1941–2006) set out to measure how the ticking of a clock is affected by a journey that any one of us could take. Between 4 October and 7 October, the intrepid scientists travelled eastwards all the way around the world on commercial airliners accompanied by four state-of-the-art atomic clocks (Figure 6.19). Then, between

Figure 6.19 Joseph Hafele (left) and Richard Keating (right) with two of their atomic clocks investigating the time-warping predictions of general relativity aboard a commercial airliner in 1971.

13 October and 17 October, they changed direction and travelled around the world westwards with their clocks. According to general relativity, time would pass at a different rate for the high-flying clocks compared to their stay-at-home counterparts—there would be contributions due to differences in velocity, as well as the gravitational effects produced by differences in altitude. The expected time disparities were then calculated using the data recorded during the circumnavigations, including air speed, altitude, latitude, and flight directions. Hafele and Keating confirmed that different periods of time had, indeed, elapsed for the airborne clocks compared to the Earth-bound clocks; the difference was tiny, but the results agreed precisely with Einstein's theory.[14]

Five years later, NASA followed up Hafele and Keating's exploits by sending a hydrogen maser clock on a two-hour flight that rose to an altitude of 10,000 kilometres before splash down in the Atlantic; it was known as Gravity Probe A and it confirmed the gravitational time dilation effects predicted by general relativity. This short and sweet mission has been overshadowed by its much more technically sophisticated cousin Gravity Probe B (GP-B). NASA launched GP-B on 20 April 2004. It had been on the drawing board for almost half a century; originally proposed in 1959 just two years after the launch of Sputnik, the world's first artificial satellite.

GP-B was launched into a polar orbit. Its objective was to use the onboard gyroscopes to probe the Earth's gravitational field. In particular, GB-P was designed to test two small effects predicted by general relativity.[15] The satellite carried four gyroscopes to improve the experimental accuracy and provide some redundancy, but only a single gyroscope is required to understand the principle of the experiment. The gyroscope is associated with a direction—the axis around which the gyroscope is spinning—this is essentially an arrow in space. In flat empty space we would expect this arrow to point in the same direction relative to the distant stars for as long as we care to look. But a gyroscope in orbit is not in empty space, it is falling freely in the Earth's gravitational field. General relativity predicts that the arrow will slowly rotate, veering in the direction of the satellite's orbital motion which, in a polar orbit, is north–south. This is due to the curvature of the space around the Earth, and it is known as the geodetic effect.[16] In the case of GP-B the direction of the arrow is predicted to shift by 0.0018 degrees per year.

There is a second even more subtle effect, first described by Josef Lense (1890–1985) and Hans Thirring (1888–1976) in 1918, that has no counterpart in Newtonian gravity. According to general relativity, a rotating massive body acts like a vortex in spacetime, dragging space around with it. This is known as frame-dragging or the Lense–Thirring effect. It is an extremely

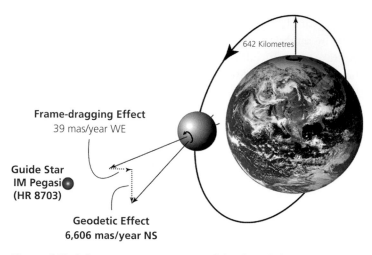

Frame-dragging Effect
39 mas/year WE

Guide Star
IM Pegasi
(HR 8703)

642 Kilometres

Geodetic Effect
6,606 mas/year NS

Figure 6.20 Schematic representation of the slow shift in the spin axis of the gyroscopes on board Gravity Probe B.

small effect in the vicinity of the Earth, but it means that the plane of a satellite's orbit will gradually shift in the direction of the Earth's rotation.[17] The gyroscopic arrows onboard GP-B are predicted to gradually shift eastwards by 39 milli-arcseconds or 0.000011 degrees per year due to this effect (Figure 6.20). In a polar orbit the geodetic effect and the Lense–Thirring effect act in perpendicular directions, making it easier to distinguish the much smaller Lense–Thirring effect.

The technical challenges of the GP-B mission were enormous, and the details are beyond the scope of our story. We can get a flavour for the sophistication of the engineering when we consider that the gyroscopes are the most perfect spheres ever manufactured. Made of fused quartz, about the size of ping pong balls, and coated with a film of niobium, they were cooled to just $2°$ above absolute zero and spun up by jets of helium. Their spin produces an electric current in the coating of niobium, which superconducts at this low temperature, and this induces a magnetic field that is monitored with superconducting quantum interference devices (SQUIDs) to determine the direction of the gyroscope's spin axis.

Although GP-B is generally regarded as a successful mission, there were unanticipated issues with the gyroscope currents that meant that the measurements did not achieve the precision originally anticipated. Nevertheless, in 2011 NASA announced that GP-B had confirmed the existence of the geodetic effect to within 1% of the predicted value and the Lense–Thirring effect to within 20%. This was the first time that the effects of the gravitational vortex had ever been seen. We now know that there are places in the depths of the universe where the consequences of such spacetime maelstroms are much more extreme, as we will see in Chapter 9.

The mismatch between the predictions of Newtonian gravity and general relativity are tiny in the vicinity of the Earth—so small that we might imagine the superior precision of general relativity could never have any practical use. It is rather surprising, therefore, that we now depend on an everyday technology—satellite navigation—that would not function without the use of general relativity. The original satellite navigation system developed by the United States is built around a constellation of twenty-four satellites in high Earth orbit, at an altitude of about 20,000 kilometres. Each satellite carries an atomic clock, and these clocks are synchronized several times a day with updates from even more accurate ground-based clocks. From any point on Earth, at least four GPS satellites are above the horizon at each moment in time. Triangulation using the time taken for signals to arrive from four satellites can be used to determine the position and altitude of any point on Earth to within two metres. In order to achieve such accuracy, the system must keep time to within about 6 nanoseconds.[18]

This means that two relativistic effects must be taken into account. The satellites travel around the Earth at about 14,000 kilometres an hour, which introduces time dilation effects, as described by special relativity which causes the atomic clocks on board to lose about 7 microseconds per day compared to Earth-based clocks. But the satellites are much higher in the Earth's gravitational field than clocks on Earth. So the orbiting clocks tick faster than clocks on the ground deeper in the Earth's gravitational well. This general relativistic effect causes the GPS satellite clocks to gain 45 microseconds per day producing a net gain of 38 microseconds per day. The system is designed to compensate for this disparity in the flow of time. If general relativity were not taken into account GPS would become unreliable in under a minute.[19]

Is Einstein's Theory Better Than Newton's?

On 31 December 1999, *Time* magazine named Einstein as the person of the century, which is quite an accolade. Einstein's greatest achievement was his theory of gravity—general relativity. So, is it really a better theory than Newton's? In every case where it has been possible to measure a gravitational effect with sufficient accuracy to distinguish between general relativity and Newtonian gravity, general relativity has always come out on top. General relativity is, without question, a more accurate theory than Newtonian gravity. Indeed, general relativity matches the current experimental and observational evidence in every situation in which it has been possible to test it. This is why it is accepted as the best theory of gravity that we have. But there is more to the theory than this; general relativity is universally regarded as the most elegant of all theories of physics. It epitomizes what physicists consider to be a beautiful theory. General relativity is constructed from a few simple philosophical principles, and this gives it very secure physical foundations.

When Newton published the *Principia*, he was criticized by philosophers and mathematicians such as Bishop George Berkeley (1685–1753), Wilhelm Leibniz (1646–1716), and Christiaan Huygens for introducing occult forces into the physical sciences.

The force of gravity seemed to act at a distance, which left a bad taste in the mouth. In Einstein's theory, this issue is resolved. According to general relativity, the presence of matter curves spacetime in its vicinity, and this local curvature then spreads outwards at the speed of light. The shape of spacetime will determine the course of other bodies, but they cannot be affected until there has been time for a gravitational influence to reach their locality.

The Schwarzschild Solution

Schwarzschild solved Einstein's equation and found the shape of spacetime around a spherical object such as a star or planet. Schwarzschild also solved Einstein's equation for the shape of spacetime within a spherically symmetrical mass. Having seen that there is extremely good evidence that Einstein's theory offers a useful way to view gravity, we should take a closer look at Schwarzschild's solutions. Space is three-dimensional and we live inside it, so we cannot view the shape of space from outside. But we can get a good idea about the shape of space if we consider a two-dimensional slice through three-dimensional space. Our surface will look like a warped rubber sheet (we must bear in mind that the passage of time is also distorted, but we will come back to that).

Outside the mass, the curvature of space is negative, so the two-dimensional slice through space looks like the hyperbolic surface that we saw earlier. The curvature decreases with distance from the mass and gradually falls towards zero. At large distances from the mass, space will be essentially flat.

Our slice through space looks like a funnel that ends on an open circle, as can be seen at the bottom of Figure 6.21. This is because, for the moment, we are only considering the space outside the mass. The circle represents a cross-section of the boundary of the spherical mass. It might be its equator, for instance. The hyperbolic funnel corresponds to the region of space outside the spherical mass where, in the Newtonian description, the force of

Figure 6.21 The geometry of a two-dimensional slice through space outside a spherical mass. The circle at the base of the diagram corresponds to where the slice passes through the surface of the spherical mass. For instance, this could be the equator of the sphere.

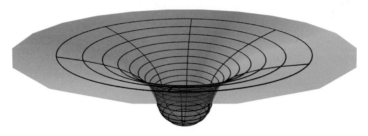

Figure 6.22 A two-dimensional slice through the space in and around a massive spherical body. Outside the body, space is negatively curved; inside the body, space is positively curved.

gravity falls off as the inverse square of distance from the centre of the mass.

We can also consider the shape of space within the mass. The direction of curvature changes as we enter the mass. Inside the mass the curvature is positive, and if the mass has uniform density, the curvature inside will be constant, and our two-dimensional slice will look like a hemisphere. The space within the mass must join smoothly to the space outside, which means that our hemisphere can be sewn on to the circle at the bottom of the previous illustration; the resulting shape is shown in Figure 6.22. To a good approximation, this is the shape of a two-dimensional slice of space in and around any spherical mass.

The spherically curved space inside the mass corresponds to the region of space where the Newtonian force increases with distance from the centre of the mass. This means that the force falls off as we approach the centre, as described in Chapter 4, where we journeyed through the centre of the Earth. At the centre of the mass, the force disappears completely. In Figure 6.22, this corresponds to the point at the bottom of the depression. As long as we remember that we are looking at a two-dimensional slice through space, this is a good picture of how space is warped by the presence of a massive object.

In reality, an object such as a star will warp the whole three-dimensional space that surrounds it. We have taken a slice through space so that we can see it embedded in flat three-dimensional space. When we do this, it is immediately obvious that the surface is curved. But imagine creatures confined within the two-dimensional surface. How would space appear to them? They cannot see the surface from outside, as we have been doing. The curvature will appear to them as a distortion in the length of their rulers. In Figures 6.21 and 6.22, the surface is marked out with equally spaced circles. We can project the surface onto a map. This will inevitably produce some distortion in the distances on the map, when compared to the real surface. The result is the map shown in Figure 6.23, which has been produced by projecting our two-dimensional slice upwards onto a flat plane.

Moving radially inwards, the circles cluster ever closer together until we reach the edge of the mass (which is situated where the circle is coloured blue)—this corresponds to the negatively curved space outside the mass. As we enter the mass, the curvature changes, and the geometry switches from hyperbolic to spherical. Now the circles gradually become more spaced out again. The circles are bunched most closely where the rubber-sheet-like two-dimensional slice plummets downwards at the edge of the mass. Creatures within the slice would see this as a distortion in the lengths of their rulers as they travel through the gravitational field of the mass. (Note that as the massive body

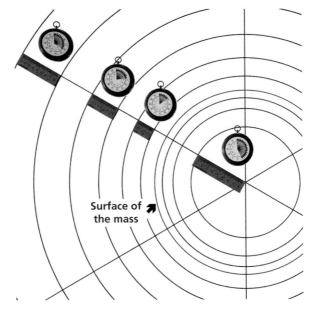

Figure 6.23 Projection of a slice through Schwarzschild space in and around an extremely dense body. The distortion in the radial direction is shown as the stretching of a standard ruler. The distortion in time is depicted as a change in the rate at which the second hand on a watch moves.

is approached rulers shrink, which means that the map under-states lengths in the actual two-dimensional surface. This corresponds to the stretching of space in the gravitational field.) Similarly, clocks slow down, as verified by Pound and Rebka, which indicates that time has been compressed in the gravitational field.

Warping Three-dimensional Space

Our two-dimensional slice is representative of all the possible slices through the space around a massive object. In each slice, space is stretched towards the mass in the same way. However, it is very difficult to draw all this information together and imagine the distortion of three-dimensional space in its entirety. The problem is that, like the imaginary two-dimensional creatures in our slice, we live within the three-dimensional space that is being warped. The best that we can do is to pretend that we live within a flat three-dimensional map and account for the warping of space from within. Like the two-dimensional creatures, we can do this by considering the effect of gravity on the length of a ruler. The curvature of the three-dimensional space in which we live can be mapped out by comparing the lengths of rulers to a standard ruler that is kept far from the mass. As shown in Figure 6.23, outside a mass, our rulers shrink in the direction towards the mass. In three dimensions, there are two directions perpendicular to the radial direction. A ruler placed in either of these directions would grow as we moved towards the mass. What this means is that space is stretched in the radial direction and squeezed in the two perpendicular directions.

Returning to the Newtonian description of gravity for a moment, the decrease in gravity with distance outside a massive object means that the gravitational force at two widely separated points may be different. The difference in the pull of the Moon on opposite sides of the Earth raises the tides and, for this reason, such differential forces are known as tidal forces. Einstein removed the need to invoke the Newtonian inverse square law force to explain gravity, but tidal forces cannot be removed in this way. In fact, it is the tidal forces that correspond to spacetime curvature in general relativity. According to Einstein's theory, the Moon distorts the shape of space in its vicinity. This spatial curvature stretches the Earth in the direction of the Moon and squeezes the Earth in the perpendicular directions. In Einstein's description it is this stretching and squeezing that generates the tides (Figure 6.24).

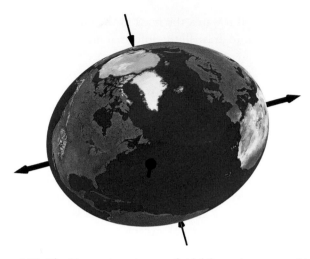

Figure 6.24 The Newtonian picture of tidal forces is converted into a picture of spacetime curvature in Einstein's theory. The tidal forces produced by the Moon's gravity correspond to a distortion of the space around the Earth, and this is what causes the tides to rise. The distortion is enormously exaggerated in the illustration. The size of the actual distortion can be gauged by comparing the height of the tides to the radius of the Earth. (The Earth is solid and so resists being stretched and squeezed by the distortion of space. The oceans are free to flow, so the distortion of space raises the tides.)

Holes in Space

The universe contains objects that are far more massive and much denser than the Sun. It is close to these objects that the predictions of general relativity really come into their own.

The curvature of space within a massive body has an effect that is quite similar to the spherical bubble in the analogy described earlier. The curved space squeezes the material within, raising its pressure and increasing its density until it exerts an outward pressure that prevents it from being compressed further. This outward pressure is produced by structural forces within the material body, such as those between atoms, that are ultimately

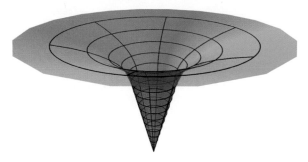

Figure 6.25 Two-dimensional slice through the Schwarzschild space around a black hole.

electromagnetic in origin. In Chapter 7, we will see that there are also mysterious stellar remnants that are supported by nuclear forces. If a star is sufficiently massive, the gravitational curvature generated by its mass may be so intense that it crushes the star out of existence. The Schwarzschild solution predicts that once a body has been squeezed within a distance known as its Schwarzschild radius, nothing can prevent further collapse; all that remains is an extremely warped region of spacetime known as a *black hole*. Such an object has a spherical boundary known as its *event horizon*, from within which not even light can escape. The Schwarzschild radius of the Sun is just 2.9 kilometres, so the entire mass of the Sun would have to be squeezed within this radius to turn it into a black hole. The Schwarzschild radius of the Earth is less than a centimetre. No known force is capable of transforming the Earth or even the Sun into a black hole, but there can no longer be any doubt that black holes really do exist. Figure 6.25 shows a two-dimensional slice through the space around a black hole. The event horizon is circular in the illustration, and spherical in the true three-dimensional picture. In Chapter 7, we will take a closer look at these celestial monsters.

7

A Brief History of Black Holes

Twinkle, twinkle, little bat!
How I wonder what you're at!
Up above the world you fly,
Like a teatray in the sky.

LEWIS CARROLL,
Alice's Adventures in Wonderland (1865) CH. 7

This is Nonsense, Stephen!

Stephen Hawking's greatest triumph came in 1974.[1] He made his breakthrough public at the Rutherford Appleton Laboratories' winter meeting, where he announced the results of his latest research to an audience of theorists. Hawking was already using a wheelchair and was on the stage. As he came to the conclusion of his talk, he broke the news of his momentous insight into fundamental physics. There was a stunned silence. The audience of experienced physicists were shocked. Finally, Professor John Taylor of King's College, London, who was acting as chairman for this session of the meeting, stood up angrily, saying, 'This is nonsense, Stephen!' before rushing from the lecture theatre— apparently intending to write a paper to demolish Hawking's idea. This is not the usual response to a fundamental physics seminar, which is more likely to end with a polite handclap and a couple of yawns from those who began to doze as the speaker conjured up the more technical details of their latest research. So what did Hawking say that was so shocking to his colleagues? I will reveal all after a quick tour of the cosmos.

Apocalypse Now!

Stars are huge writhing balls of hydrogen and helium. Despite their almost identical composition, their ultimate fates can be remarkably different. The real action within a star takes place in its core, where conditions are so extreme that atomic nuclei merge to form heavier nuclei. The energy released in this way supports the star against the crush of its own gravity, but eventually the fuel runs out and the nuclear reactor in the star's core is turned off and, once the energy supply is exhausted, the star may collapse under its own gravity, as we will see.

When a star has converted most of the hydrogen in its core into helium, its outer layers swell up to form a bloated red giant. Five billion years from now, the Earth will be engulfed by the outer layers of the Sun as it approaches the end of its life. In a star such as the Sun, these outer layers will eventually disperse into space to reveal the star's core as an extremely dense glowing ember, about the size of the Earth but with the mass of a star (Figure 7.1). The nuclear reactions in the core have now ceased, so the core gradually cools as it radiates its heat into the depths of space. Cosmic cinders, such as these, are called white dwarfs.[2] They are extremely hot, but very faint, as they are so small compared to normal stars. The nearest white dwarf is in orbit with Sirius, the brightest star in the night sky, but it is not visible without a large telescope.[3]

In 1930, Subrahmanyan Chandrasekhar (1910–1995) was awarded a graduate scholarship by the Indian government to continue his studies in physics at Trinity College, Cambridge. While on the voyage to England, Chandrasekhar realized that there is an upper limit to how much mass could be supported in the form of a white dwarf and calculated that it is just under one and a half times the mass of the Sun. This mass is known as the Chandrasekhar limit. According to Chandrasekhar's calculations any white dwarf with a mass greater than this will collapse under its own gravity. But what happens then? What

Figure 7.1 Size comparisons of four celestial bodies. From left to right: A white dwarf, which is the same size as the Earth; Jupiter; the Sun; the red giant Aldebaran, which is slightly more massive than the Sun.

happens to matter when it is crushed beyond the density of a white dwarf? This is not simply an idle question; many stars are much more massive than the Sun, so what happens when they run out of nuclear fuel?

Although the Sun will end its days as a white dwarf, a more spectacular future awaits the really massive megastars. Stars of greater mass have higher temperatures in their core, which means that the nuclear reactions proceed faster and progress further. Very massive stars burn their nuclear fuel at a prodigious rate and race through their life much quicker than lower-mass stars. The Sun is about half-way through its 10 billion-year life span. A megastar with around twenty times the Sun's mass will exhaust its nuclear fuel in 10 million years. Twenty times as much fuel is burnt in one-thousandth of the time, so the energy is released 20,000 times faster and consequently the star shines 20,000 times brighter than the Sun. Fortunately, these stars are all much further away than the Sun, so we only see them as tiny pinpricks in the night sky.

For most of a star's life, it is supported against gravitational collapse by the thermal energy generated by fusing hydrogen into helium in its core. When the hydrogen runs out, the core contracts and the temperature rises. In higher mass stars this triggers new rounds of nuclear fusion processes. First, helium is converted into carbon and oxygen, then heavier atoms are cooked up; neon and magnesium; then sulphur and silicon; and finally iron and nickel. Iron is the densest nucleus, so at this point no further fusion energy can be extracted. There is now no source of energy to balance the tendency of the star to contract under gravity. As the core collapses, huge amounts of gravitational binding energy are released and the star blasts itself apart in a supernova explosion—a cosmic firework that may be as bright as an entire galaxy composed of hundreds of billions of stars. But these explosions, which are known as Type II supernovae, are not the only way in which a star can blast itself apart.

A Ticking Time Bomb

Many stars live in binary or multiple star systems, in which two or more stars are bound together and orbit each other. If a white dwarf and a red giant are held in a close gravitational embrace, their interactions can be dramatic. The white dwarf is, essentially, a dead star, as its internal nuclear fusion reactions have ceased. But as the white dwarf and the red giant pirouette, the white dwarf may draw hydrogen and helium from the outer layers of its swollen partner. This material builds up on the white dwarf's surface to form a dense shell, compressed by the white dwarf's intense gravity. Eventually, a critical mass accumulates and the shell detonates in a huge nuclear fusion explosion that is visible from the other side of the galaxy.

Such events are regularly seen by astronomers. Suddenly a star appears as if from nowhere. It is known as a nova, meaning 'new star'. Gradually, the nova fades and eventually it disappears again. About ten novae are seen in the Milky Way galaxy each year (another thirty or so are thought to be hidden from view by dust and gas clouds). The process leading to the nova will repeat as the white dwarf continues to draw material from its companion. Typically, the period between eruptions is several thousand years, but it may be as short as a decade or two. For example, the star RS Ophiuchi lit up in 1898, 1907, 1933, 1945, 1958, 1967, 1985, 2006, and most recently in August 2021.

Over time, a white dwarf's mass may increase until it approaches the critical Chandrasekhar limit—the point at which the white dwarf can no longer resist collapsing under its own gravity. This triggers a runaway fusion explosion that consumes the entire white dwarf producing a cataclysmic conflagration that may be 100,000 times brighter than a nova—the star has gone supernova. These explosions are known as Type Ia supernovae (Figure 7.2). The precise details of how this occurs are still being researched, but it seems that in an instant much of the white dwarf is transformed into heavy elements and blasted at high speed into the depths of space. Type Ia supernovae are extremely

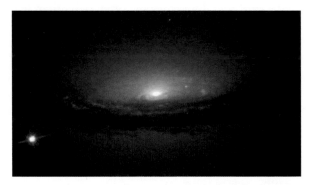

Figure 7.2 The bright star in the bottom left of the picture is a type Ia supernova that erupted in 1994 in the outskirts of a galaxy known as NGC 4526, which is 100 million light years away. It was comparable in brightness to the core of the galaxy which contains around 100 billion stars.

bright and all have a similar luminosity as the detonating white dwarfs all have the same mass. It was the sudden appearance of a Type Ia supernova where no star was previously seen that astounded Tycho in 1572.

We have seen two distinct mechanisms that produce supernova explosions. Very massive stars that have used up all their nuclear fuel end their lives with a bang. Also, greedy white dwarfs may accumulate so much extra mass that they undergo a runaway thermonuclear fusion explosion. The two mechanisms are quite different, and the resulting supernovae can be distinguished by astronomers. They were classified as Type Ia and Type II before their origins were understood.

So what is left of the star when the smoke clears?

Little Green Men

It is an interesting question—if one thinks one may have detected life else-where in the universe how does one announce the results responsibly? Who does one tell first?

Jocelyn Bell Burnell, *Little Green Men, White Dwarfs or Pulsars?*[4]

Towards the end of 1967, Jocelyn Bell (1943–) was analysing the signals detected by a new array of radio receivers constructed in Cambridge, under the guidance of her PhD supervisor Antony Hewish (1924–). Bell noticed a signal within the data that repeated with metronomic regularity every 1.337 seconds.[5] Such regularity had no obvious explanation. After ruling out a terrestrial origin for the radio pulses, another explanation was that the Cambridge astronomers had found the first sign of an extraterrestrial civilization, and they seriously considered this possibility. The radio source was whimsically designated LGM-1, where LGM stands for 'Little Green Men'.

On Christmas Eve, Bell found another example pulsing away in a different part of the sky, and this was followed by two more examples after the Christmas holiday (Figure 7.3). Clearly, the sky could not be full of extraterrestrials that were all attempting to attract our attention in the same way. The true identity of this

Figure 7.3 Jocelyn Bell as a graduate student.

strange new class of celestial objects, now known as pulsars, was provided by the astrophysicist Tommy Gold (1920–2004) who suggested that pulsars must be the tell-tale signs of *neutron stars*. His arguments were persuasive and were soon accepted by the physics community.

A neutron star forms in a supernova explosion when the collapsing core of a star is squeezed beyond the density of a white dwarf and transformed into a sphere that is just 25 kilometres in diameter—the size of a major city—but with the density of an atomic nucleus. These outlandish ultra-dense objects spin at an incredible rate, typically completing a full rotation in a fraction of a second. This generates a huge magnetic field that causes the emission of two intense beams of radiation from the neutron star's magnetic poles. Pulsars are like cosmic lighthouses. The pulsar beams are not perfectly aligned with the neutron star's rotation axis so as the star spins they sweep across the heavens. On Earth, radio astronomers detect a pulse of radio waves once every rotation when the beam points in our direction, which may be many times a second.

The connection between supernovae and neutron stars was clinched by the discovery of a pulsar in the Crab nebula (Figure 7.4), which has been identified as the remnants of a

Figure 7.4 The Crab nebula.

supernova seen in the year 1054 and recorded in the ancient Chinese annals. Within the Crab nebula is a neutron star that spins thirty times every second and sends a pulse of radio waves in our direction once every rotation. This is the collapsed core of the star that blasted itself apart in the supernova explosion.

White dwarfs have a mass comparable to the Sun compressed into a sphere the size of the Earth, as depicted in Figure 7.1. If a teaspoonful of white dwarf material were returned to Earth it would weigh a staggering 10 tonnes. But neutron stars are in another league entirely. The mass of a neutron star is about one and a half times that of the Sun and it is compressed into a sphere whose diameter is 25 kilometres. On Earth, a teaspoonful of neutron star would weigh at least 1 billion tonnes.

There is a maximum mass that can be supported as a neutron star but the physics of these weird objects is so exotic that physicists cannot determine this limit with the same accuracy as for white dwarfs. Nonetheless, we know for sure that the maximum mass of a neutron star is about two and half solar masses, and certainly less than three. So, what happens to the collapsing core of a star whose mass is greater than this?

A One-way Trip to Oblivion

In 1939, Robert Oppenheimer (1904–1967) and his research student Hartland Snyder (1913–1962) considered the ultimate fate of a collapsing ball of matter and what Einstein's theory of general relativity has to say about it. They concluded that after exhausting its nuclear fuel, a sufficiently massive star must undergo total collapse to produce an ultra-dense body from which neither matter nor light could ever escape. This is the modern incarnation of John Michell's idea of a dark star that we met in Chapter 4. Oppenheimer and Snyder concluded:

The star thus tends to close itself off from any communication with a distant observer; only its gravitational field persists.

Many years later the American physicist John Archibald Wheeler (1911–2008) found an appropriate name for these bizarre objects that are so dense that even light is trapped within. Wheeler named them *black holes*. To escape from within a black hole, a particle would have to travel faster than the speed of light, and one of the fundamental assumptions of relativity is that this is impossible.[6] But can such objects really exist? Even Einstein had his doubts. He felt that some unknown laws of physics would come into play in the extreme conditions necessary to form a black hole and that these laws would prevent the creation of such outlandish objects.

Einstein, who died in 1955, did not live to see this question resolved. Oppenheimer and Snyder's analysis was based on what would happen when a spherically symmetrical mass collapses. Other theorists were concerned that they had simply considered an idealized and unrealistic model that did not reflect reality, arguing that the creation of a black hole in the Oppenheimer–Snyder model was an artefact of their assumption of perfect spherical symmetry. After all, if all the points of mass in a perfect sphere collapse inwards under gravity, it is inevitable that they will meet in a point of infinite density at the centre of the sphere. Was this just a geometrical consequence of the initial premise of spherical symmetry or did the Oppenheimer–Snyder model actually show what would happen in the real world? Russian physicists Evgeny Lifshitz (1915–1985) and Isaak Khalatnikov (1919–2021) were sceptical. And, in a paper published in 1963, they claimed they had a proof that a black hole would not form in the more natural situation of a messy asymmetrical collapse, casting serious doubt on the existence of such objects. John Wheeler brought this issue to the attention of Roger Penrose (1931–) who was visiting the Princeton Institute of Advanced Study as a young graduate student. Wheeler posed the question: in the general case where no symmetry is assumed would the gravitational collapse of a star result in the creation of a black hole?

Initially, Penrose (Figure 7.5) could not see a way to characterize the problem of asymmetrical collapse. This all changed one day

Figure 7.5 Sir Roger Penrose.
Credit: © Nobel Prize Outreach. Photo: Fergus Kennedy.

when he took a walk with a colleague Ivor Robinson (1923–2016) who was visiting Princeton. According to Wolfgang Rindler, '[n]o one who knew Robinson will forget what a brilliant conversationalist he was, with his sonorous deep voice and ultra-English accent'.[7] Indeed, Robinson was a rather garrulous British physicist and during his walk with Penrose he only stopped talking when they came to a busy road. As soon as they had crossed, Robinson resumed his discourse. Later that day, Penrose had an odd sense of elation, but he could not put his finger on why. So he retraced his thoughts through all the events of the day. Then eventually he recalled crossing the road with Robinson and realized that just at that moment he had thought of an idea that was key to understanding the formation of black holes. It was the idea of a trapped surface.

In flat space, the light rays emitted by a surface will generally diverge, parting forever on their way out into the depths of space. Penrose defined a trapped surface to be a closed two-dimensional surface with the property that all light rays emitted from the surface converge. This is precisely the situation within a black hole's event horizon where the intense gravity focuses all light rays inwards where they meet within a finite period of time. (Even outgoing light rays cannot escape from a black hole, they are trapped within and are directed by the extreme spacetime curvature to the centre of the black hole.) Penrose showed that the existence of a trapped surface always results in an unstoppable collapse. This allowed him to conclude that the gravitational collapse of a sufficiently massive star must inevitably lead to the creation of a black hole. Furthermore, according to general relativity, any matter within the black hole collapses into a singularity, or point of infinite density. In short, a black hole is defined by its event horizon. Once inside, nothing can escape, not even light. Stray across the boundary and you are on a one-way trip to oblivion.

Penrose visited London soon after his road-crossing epiphany and gave a seminar about his idea of trapped surfaces. This seminar features in the film *The Theory of Everything*, the life story of Stephen Hawking, with its remarkable portrayal of Hawking by Eddie Redmayne. In reality, Hawking was not at this particular seminar, but Penrose agreed to give a follow-up seminar in Cambridge, and this time Hawking was present. Following the seminar Hawking worked with Penrose and adapted his ideas to the very early universe. By running time backwards Hawking showed the inevitability of a singularity at the dawn of the universe—the Big Bang.

Although Penrose had put the theoretical basis for black holes on a firm footing, their existence in the real world remained an open question. Could such bizarre objects actually exist or did the laws of nature conspire to somehow prevent their creation? If astronomers were to find evidence for black holes, they would need to know what to look for.

Following the resurgence of interest in black holes during the 1960s, Edwin Salpeter (1924–2008) and Yakov Zeldovich (1914–1987) considered how they might be tracked down in the depths of the universe. They argued that a black hole would be surrounded by a rapidly rotating disc known as an *accretion disc*, formed of gas and dust accumulated from its neighbourhood. As the material in this disc falls towards the black hole its gravitational binding energy is released and its temperature rises to millions of degrees. Indeed, the accretion disc would be so hot that it would emit X-rays. Salpeter and Zeldovich concluded that X-ray emission would be the tell-tale sign of a black hole.

What Would a Black Hole Look Like?

Black holes have played starring roles in science fiction films. Unfortunately, many depictions of black holes are rather misleading. Black holes are sometimes represented as giant plug holes in the sky, as shown on a 50-pence coin minted to commemorate the life of Stephen Hawking (Figure 7.6). This is completely incorrect.

Black holes definitely should not be thought of like giant sinkholes in the ground—a black hole is a spherical region of space. So what do black holes look like? The planet Saturn offers a surprising analogy for their geometry (Figure 7.7). There are obvious differences, but first we will consider the similarities.

The event horizon of a black hole is spherical, so like Saturn, black holes are spherical. Once inside the event horizon there is no escape, even for light. But much of the material in the vicinity

Figure 7.6 Stephen Hawking commemorative 50p piece.
Credit: Copyright: The Royal Mint, 2019.

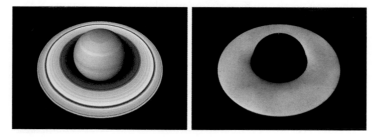

Figure 7.7 Left: Saturn, Right: Simulated black hole.

of the black hole, such as rock or dust or gas, accumulates in an accretion disc, as Salpeter and Zeldovich suggested, and this material swirls around the black hole in the plane of the black hole's equator. This disc is reminiscent of Saturn's rings. Between the event horizon of the black hole and the inner edge of the accretion disc there is a gap, just like the gap between Saturn and its rings.

Now we come to the differences. The gravitational field of a black hole is so intense that nothing can escape, not even light, so the sphere that defines the black hole is completely black. Furthermore, whereas Saturn's majestic rings perform a serene orbital dance around the planet, the black hole's accretion disc orbits at a ferocious pace approaching the speed of light. The accretion disc is composed of material that fell towards the black hole releasing huge amounts of gravitational binding energy, so it is hot—very hot. It is a violently swirling cloud of plasma composed of high-temperature material whose atoms have dissociated into electrons and atomic nuclei. The radiation emitted by this blazing hot plasma would indeed prove to be the key to finding conclusive evidence for black holes.

The X-ray Is Her Siren Song

In 1970, NASA placed its first X-ray telescope in Earth orbit. It blasted off from a launch site in Kenya on 12 December, the seventh anniversary of Kenya's independence from Britain. NASA named the satellite *Uhuru*, Swahili for freedom, in honour of their Kenyan hosts.[8] The telescope was designed to explore space in a new region of the electromagnetic spectrum, viewing it as it had never been seen before. X-rays correspond to electromagnetic radiation of short wavelength and high energy. They are only produced in violent events in very energetic environments. (Fortunately, we are shielded by the Earth's atmosphere from high-energy radiation from space, which is why X-ray astronomy is only possible from Earth orbit.) Like any pioneering journey into the unknown, when the satellite was launched, no one was sure what it might find.

One of the targets for investigation by Uhuru was a powerful X-ray source in the constellation of Cygnus the Swan that had been discovered in 1964. This object is known as Cygnus X-1. It would turn out to be a very interesting celestial object. The X-ray source was soon identified with a star that could be examined in visible light with earth-based telescopes. Astronomers have now been studying Cygnus X-1 for over fifty years with a variety of observatories, detecting light at a range of wavelengths—radio, optical, and X-ray. The best data come from NASA's X-ray satellite Chandra, launched in 1999 and named after the Indian astrophysicist Chandrasekhar who was awarded the Nobel Prize in Physics in 1983 for his research into the physics and evolution of stars. Assessment of the most up-to-date information has enabled astrophysicists to draw up a detailed picture of how the X-rays are produced.

Astronomers believe that Cygnus X-1 was once a binary star system consisting of two very massive stars bound together by their mutual gravitational attraction. The heavier of the two was a mighty star of sixty solar masses that exhausted its nuclear fuel first and, in the subsequent supernova explosion, its core collapsed to form a black hole. The original star would have been millions of kilometres in diameter, but the black hole is tiny, with a diameter of only a few tens of kilometres (less than a millionth of its original diameter). The black hole remains gravitationally bound to its companion star, and they continue to orbit each other. Cygnus X-1 has been monitored by the Very Long Baseline Array, which consists of ten radio telescopes spread across the United States, and its distance is estimated as 7240 light years. The latest data suggests that the black hole's mass is just over twenty times that of the Sun.[9]

The black hole's companion star is a brilliant blue supergiant that can easily be seen with a modest telescope. The black hole and the supergiant dance around each other in a tight embrace, with each orbit taking just over five and a half days. The outer layers of the star are dispersing into space. Some of this plasma accumulates around the black hole, to form an accretion disc rotating in the plane of the black hole's equator. There are no stable orbits close to the black hole so the plasma forming the inner edge of the accretion disc gradually spirals into the black hole; a huge amount of gravitational energy is released as the material falls inwards and this is converted into heat by friction in the swirling disc. The accretion disc glows so hot that it emits X-rays, and this was the distinctive clue that revealed that Cygnus X-1 is home to a black hole, as only in the most extreme environments is matter heated to the enormous temperatures required for X-ray emission.

The radius of the event horizon of a non-rotating black hole is equal to its Schwarzschild radius, which is given by the formula

$$r_S = \frac{2GM}{c^2},$$

where r_S is the Schwarzschild radius, G is Newton's constant, M is the mass of the black hole, and c is the speed of light. As mentioned in the previous chapter, the Schwarzschild radius of the Sun is just under 3 kilometres. If the entire mass of the Sun were squeezed into a sphere of this radius, then it would become a black hole. The Cygnus X-1 black hole is about twenty times as massive as the Sun. Its Schwarzschild radius is therefore twenty times that of the Sun—60 kilometres. And if it were not rotating this would be the radius of its event horizon. In fact, the black hole is spinning at a phenomenal 800 times per second—close to the maximum rate possible—and this reduces the size of its event horizon. The radius of the event horizon of a black hole spinning at the maximum possible rate is half its Schwarzschild radius.

So black holes are tiny by cosmic standards. The diameter of the Cygnus X-1 black hole is probably much less than the thickness of its accretion disc. Intense magnetic fields are generated in the accretion disc, and in this highly energetic and turbulent environment, not all of the plasma finds its way into the black hole. Some is propelled outwards in the form of two oppositely aligned jets that spew hot matter into the depths of space from near the poles of the black hole. An artist's impression of the Cygnus X-1 system is shown in Figure 7.8.

Figure 7.8 Artist's impression of the Cygnus X-1 system. Material from the blue giant star is being drawn towards the black hole to form an accretion disc swirling around the black hole. There is a bright spot where this material hits the accretion disc.

A Black Hole Has No Hair!

In 1963, Roy Kerr (1934–), a mathematician from New Zealand, discovered a remarkable solution of Einstein's equation. It describes the shape of spacetime around a rotating spherical mass. For a non-rotating mass, the Schwarzschild solution is identical to Kerr's solution. But in the real world we can expect dense massive bodies to be spinning very rapidly. The Kerr solution is particularly important for describing the spacetime around a black hole, as black holes are the most extreme gravitational environments and are expected to spin at a dizzying pace.

In a lecture in 1975, Chandrasekhar told his audience that in his entire scientific life: 'The most shattering experience has been the realization that an exact solution of Einstein's equations of general relativity, discovered by the New Zealand mathematician Roy Kerr, provides the absolutely exact representation of untold numbers of massive black holes that populate the universe.'[10] Kerr's discovery has many important consequences. It began a golden age in the analysis of general relativity and black holes.

Planets come in all shapes and sizes. Each of the eight planets in the solar system has its own particular character and the same is no doubt true of the many others that are being found in planetary systems throughout the galaxy. Similarly, stars come in numerous varieties—red giants, brown dwarfs, neutron stars, white dwarfs—and each has a slightly different composition. Some contain more carbon than average, some contain more of various other elements—each star has a unique composition. Black holes, by contrast, have at most *three* attributes, so they are completely defined by just three numbers: their mass, the rate at which they spin, and their electric charge. Of these, the third is unlikely to be important in the real universe, as matter is usually electrically neutral and there is no known mechanism for giving a star or black hole a significant electric charge. This means that a real black hole is almost totally devoid of features; it can be described *exactly* by its mass and its rate of spin. It can have no other attributes whatsoever. Wheeler had a knack for capturing the essence of a physical idea with a short pithy statement. He referred to this result as the *no-hair theorem*, because, in his words, 'black holes have no hair'.

Hawking's Area Theorem

The radius of the event horizon of a black hole is determined by its mass—the greater the mass, the bigger the hole. As material falls into a black hole, its mass increases and, therefore, its size also increases. Nothing can get out of the black hole so, as time passes and the black hole feeds on the dust and plasma in its surroundings, it will inevitably grow in size. One way of putting this is to say that the area of the event horizon of a black hole will inevitably increase with time.

The collision and merger of two black holes to form a single black hole is an extremely violent occurrence, and we know that such collisions do happen. We will take a look at their consequences in the next chapter. Hawking noticed an important consequence of such mergers. He proved that whatever the details of the encounter, the area of the event horizon of the final black hole would be greater than the sum of the areas of the event horizons of the two original black holes. Hawking then showed that in any process whatsoever, according to general relativity, the total area of black hole event horizons must always increase. This is Hawking's Area Theorem. It is an abstract result that Hawking proved using very sophisticated mathematical techniques. It sounds like a quirky fact without much physical significance but, as we will see, it opened the door to a profound re-evaluation of the role of gravity in the universe.

The Two Cultures

In 1959, the scientist and author CP Snow (1905–1980) gave the Rede Lecture at the Senate House in Cambridge. It was subsequently published under the title *The Two Cultures and the Scientific Revolution*. In the lecture, Snow decried the gap that had opened up between scientists and artists in the twentieth century. Harking back to earlier centuries, Snow observed that science and art were considered to be two complementary ways to view the world and that an educated person was expected to be conversant with both:

> A good many times I have been present at gatherings of people who, by the standards of the traditional culture, are thought highly educated and who have with considerable gusto been expressing their incredulity at the illiteracy of scientists. Once or twice I have been provoked and have asked the company how many of them could describe the Second Law of Thermodynamics. The response was cold: it was also negative. Yet I was asking

something which is the scientific equivalent of: Have you read a work of Shakespeare's?[11]

So what is the Second Law that CP Snow was so concerned about?

Brewing Up New Theories of Physics

Physics and technology have long had a mutually stimulating effect on each other. The two are so closely intertwined that it is often difficult to disentangle the direction of these influences. One branch of physics that certainly owes a lot to technology is thermodynamics, a subject that grew out of the analysis of steam engines and how to improve their efficiency. As we demonstrate every day in a variety of ways, from boiling an egg to driving a car, energy can be converted from one form into another, and this is what thermodynamics is all about.

James Prescott Joule (1818–1889) was the son of a wealthy Manchester brewer who was taught physics by John Dalton (1766–1844), famous for introducing the idea of atoms into chemistry. Joule's own research began with experiments to see whether he could improve the efficiency of the brewery. His investigations led to one of the most important principles in physics—the First Law of Thermodynamics—which he established in the 1840s (Figure 7.9, Left).

Although energy may be converted from one form into another, the First Law of Thermodynamics states that if all forms of energy are taken into account, including heat, then the total amount of energy never changes—the sum total of energy at the start of a process will always equal the sum total of the energy at the end of the process.[12] This principle, also known as the Law of Conservation of Energy, clearly has important industrial applications; we may burn gas to heat water and produce steam that drives a turbine to generate electricity that powers our computers, but at every step, if we include the heat that may be lost along the way, the total amount of energy remains the same.

Figure 7.9 Left: James Prescott Joule. Right: Sadi Carnot aged seventeen painted by Louis-Léopold Boilly in 1813.

Another important consequence is that energy cannot be created out of nothing, it can only be changed from one form into another. Following Einstein's relativistic insights in the early years of the twentieth century, as discussed in Chapter 5, the Law of Conservation of Energy has been updated to include the fact that mass and energy are interconvertible. In effect, mass is just another form of energy. Our standard unit of energy is known today as the joule in honour of James Prescott Joule and his work.

Anyone for Iced Tea?

In the early decades of the nineteenth century the development of steam power in Britain was rapidly increasing her industrial and military might. The French engineer Sadi Carnot (1796–1832) realized that if France was going to maintain her military standing she would need more efficient steam engines. Carnot had trained as an army engineer and was the son of the Minister for War, Lazare Carnot, who had established France's

Revolutionary Army. Sadi Carnot is responsible for the second
pillar of thermodynamics (Figure 7.9, Right).

Many processes conserve energy but are never seen. For instance, heat never spontaneously flows from a cold object, such
as a lump of ice, into a warm object, such as a cup of tea—we cannot boil a cup of tea by adding any amount of ice to it. But why is
it not possible to heat up your cup of tea by putting ice cubes in
it? This suggestion is so obviously ridiculous that it takes a bit of
thought to work out what the question actually means. Our lifelong experience of interacting with the world tells us that putting
ice in our tea always cools the tea down and never warms it up.
Yet, the ice cubes that we take from our freezer contain plenty of
heat—they are much warmer than a winter's day in Siberia, for
instance. Perhaps some of this heat could be liberated from the ice
and added to the tea, thereby warming the tea and cooling the ice
further. Although this would conserve energy, we know that it
never happens.

Carnot invented the concept of *entropy* as a book-keeping device
to account for this prohibition. He defined entropy as a quantity of heat divided by the temperature. So, for a given amount
of heat, the entropy is smaller at a high temperature than at a
low temperature. This means that when heat is transferred from
a hot object to a cold object, entropy increases, whereas if heat
were transferred from a cold object to a hot object entropy would
decrease. We are very familiar with the first of these processes,
but our teatime experiments tell us that the second process never
happens. Carnot captured this observation in the statement that
in any allowed process the total entropy of the universe can never
decrease.[13] This is the Second Law of Thermodynamics.

Carnot provided a succinct prescription that distinguishes between allowed processes and forbidden processes, but he did not
explain why these processes were prohibited. They were simply
forbidden by fiat. Today, the Second Law of Thermodynamics is
explained through the statistical behaviour of large numbers of
interacting particles.

Although entropy is a more abstract concept than energy it plays an equally important role in determining which processes happen and which are forbidden by the laws of physics. One consequence of the Second Law is that when energy is converted from one form to another in a power station, a portion of the energy is always emitted as heat. It is therefore impossible to convert all the energy in a lump of coal into electricity. Another consequence is that by increasing the temperature difference between the turbines and the surrounding environment, the proportion of the energy that is lost as heat can be reduced, thereby improving the efficiency of a power station (Figure 7.10). So there are very important real world applications of the Second Law, as well as some rather otherworldly applications.

Physicists have established many conservation laws; the conservation of energy and the conservation of electric charge are just two examples. One unusual feature of the Second Law of Thermodynamics is that entropy is not conserved—it can

Figure 7.10 Battersea Power Station on the banks of the Thames in London. (Famous for its flying pigs.)

increase, but it can never decrease. This means that changes in entropy determine the direction in which a process will occur. By contrast, take energy conservation. If state A has the same energy as state B, then state A might evolve into state B or vice versa; the law of energy conservation allows either. But entropy increase is a one-way street; it gives a direction to processes that take place within our universe. The Second Law determines how the universe will evolve. It is closely related to the passage of time and our perception of the direction of time.

Black Hole Dynamics

You might be wondering what a section about thermodynamics is doing in a book about gravitation and a chapter about black holes. Remarkably, there is a very deep and subtle connection between thermodynamics and black holes, and it came as a huge surprise to physicists.

Everyone worries from time to time, and some cares are bigger than others. There are physicists who worry about the entire universe. Jacob Bekenstein (1947–2015) was one of those. Bekenstein (Figure 7.11) was born in Mexico City but grew up in the United States, and during his physics career he worked in the United States and Israel. In the early 1970s, Bekenstein was concerned that black holes might offer the universe the ultimate waste-disposal system by which it could lose some of its entropy. Black holes are featureless and seemingly have no entropy, so any material falling into a black hole would be lost to the rest of the universe, along with its entropy. Any matter that crosses a black hole's event horizon is on a one-way trip, it can never escape again. But if a black hole has zero entropy, then the entropy of this matter can no longer contribute to the total entropy of the universe. Therefore, the entropy of the universe could be reduced simply by dropping stuff into a black hole, which would violate the Second Law of Thermodynamics. This was believed to be impossible, so it was all very puzzling.

Figure 7.11 Jacob D Bekenstein in his office at the Hebrew University in Jerusalem (2009).

After some deep contemplation, Bekenstein realized that there might be a solution. When matter falls into a black hole, the black hole's mass increases and it therefore increases in size; in particular, the surface area of the black hole's event horizon increases. Perhaps the increase in the surface area of the event horizon might compensate for the loss of entropy contained in the material that had fallen into the black hole. Hawking had recently proved his area theorem which implies that in all physical processes the total area of black hole event horizons can never decrease. Bekenstein was struck by the parallel between Hawking's result and the Second Law of Thermodynamics. He proposed a generalization of the law by combining:

Hawking's Area Theorem: *The total surface area of all the black holes in the universe can never decrease*

and the Second Law of Thermodynamics: *The total entropy of the universe can never decrease.*

to produce the generalized Second Law of Thermodynamics:
The total entropy of the universe outside event horizons of black holes, plus the total area of all the black hole event horizons, can never decrease.

This is rather a mouthful, but it can be made much more succinct simply by identifying the area of a black hole's event horizon as a measure of the black hole's entropy. This is what Bekenstein tentatively proposed. But how could this be? The area theorem of black holes is a result about gravity and geometry, whereas the Second Law of Thermodynamics is a statistical law about heat.

A Restless Night

Now this was all very strange. How could there be a link between gravity and a theory that was devised to explain steam engines? Furthermore, black holes were believed to be essentially featureless, characterized simply by their mass, the rate at which they spin, and possibly their electric charge. How could they have any statistical properties?

Hawking's reaction when he heard of Bekenstein's proposal was that it was ridiculous. It could not possibly be true. He went to bed that night convinced there must be a fundamental flaw in the argument. But he couldn't sleep; he lay awake looking for a counterargument that would undermine this absurd idea. It was clear to Hawking that if a black hole had entropy, then it must also have a temperature and that was nonsense. All objects with a temperature greater than absolute zero emit radiation, and the hotter the object, the greater the intensity of the radiation. But black holes are inherently black—they cannot emit radiation because nothing, not even light can escape a black hole.

Then Hawking had a revelation. He realized there was a fatal flaw in his own reasoning and that Bekenstein must be right—the surface area of a black hole really does measure the black hole's entropy. Indeed, contrary to what everyone believed a black hole must behave like a body with a well-defined temperature and it must therefore emit radiation. Despite this counterintuitive conclusion, everything fitted together perfectly. There was no chance that Hawking would sleep now. Hawking's illness had progressed to the point where he could no longer get out of bed by himself, so he had to wait several hours for his nurse to arrive before he could put his thoughts down on paper.

Hawking Radiation

General relativity remains the best theory of gravity that we have. But general relativity is a classical theory not a quantum theory and this was the crux of Hawking's idea. It is true that when a black hole is modelled using general relativity its temperature is zero, because, as everybody knew, radiation may fall into a black hole but nothing ever comes out, so black holes do not give off heat. According to general relativity, a black hole is completely black. But although general relativity is a fantastically accurate theory, the world plays by the rules of quantum mechanics, so the ultimate theory of gravity must include features of both general relativity and quantum mechanics, and a complete description of black holes in the real world must incorporate quantum effects. Hawking realized that when these quantum effects are included, a black hole will have a non-zero temperature, and this means that a black hole must emit radiation. This was the shocking announcement that Hawking made to his colleagues in February 1974.

Hawking had recognized the first connection between gravity and quantum mechanics that anyone had seen, so it is no surprise that physicists took a while to realize that his bold idea must be true. It is very surprising that a temperature can be assigned to a black hole and contrary to what everyone expected. The radiation emitted by a black hole is now known as Hawking radiation.

If you fell into a black hole you would be crushed and stretched and it would get pretty warm. In fact, you would be vaporized well before reaching the centre of the black hole. But the temperature of the black hole referred to by Hawking is the temperature measured by someone outside the black hole. This is determined by the radiation or, in other words, heat that the black hole is emitting. So, despite whatever incredible violence might be going on inside the black hole its temperature can be very low. A black hole with the mass of a star has a temperature that is immeasurably low. Hawking's calculations show that the temperature of a black hole of ten solar masses is less than one ten-millionth of a degree (10^{-7} K) above absolute zero.[14] Any radiation emitted by such a black hole would be completely swamped by background radiation, so unfortunately it would be completely undetectable.

The temperature of a black hole is so low that it would inevitably absorb more radiation than it would emit. The universe is bathed in radiation that was produced shortly after the Big Bang, as we will see in Chapter 9. This radiation is known as the cosmic microwave background. It has a temperature that is very low—at around 2.7° above absolute zero—but it is very high compared to the Hawking temperature of a black hole. Any black hole with a lower temperature than the microwave background must necessarily absorb more radiation than it emits.

Mini Black Holes

One way to see the connection between the size of a black hole and its temperature is to consider the wavelength of the radiation that it emits. According to Hawking's quantum analysis, black holes only emit radiation with a longer wavelength than the size of the black hole's event horizon. Any radiation with a shorter wavelength than this is confined within the black hole, but electromagnetic radiation of longer wavelength seeps out as the black hole is not big enough to contain such large waves. This radiation is generated by random fluctuations of the electromagnetic field within the black hole, and its escape is a form of quantum tunnelling. The diameter of the event horizon of a stellar mass black hole is at least several kilometres and as longer wavelength corresponds to lower energy, the radiation emitted by these black holes carries very little energy. The temperature of a body that only emits radiation with a wavelength several kilometres long must be incredibly low—just above absolute zero, and this is why the temperature assigned to a stellar mass black hole is a tiny fraction of a degree above absolute zero.

On the other hand, a black hole with an event horizon just a few hundred nanometres across would radiate visible light and its temperature would be correspondingly higher—typical of objects that are so hot they visibly glow. There is no known mechanism to produce such mini black holes. Nevertheless, Hawking speculated that if the early universe was quite lumpy, then some of the denser regions might have undergone the ultimate gravitational collapse to form *primordial* mini black holes in the immediate aftermath of the Big Bang. These hypothetical mini black holes might have a mass equal to that of an asteroid packed into a region smaller than an atom. Their relatively low mass would make them much hotter than stellar mass black holes and, being hot, they would emit large amounts of radiation, thereby losing some of their mass. This decrease in mass would raise their temperature, and this would further increase the amount of radiation they emit in a runaway process that can only end one way. The temperature of a mini black hole would rise dramatically in its final moments until it exploded and disappeared in a huge blast of radiation.

If primordial black holes really do exist, then it should be possible for astronomers to detect them going bang as they disappear in a puff of gamma rays. The smallest such primordial black holes would already have exploded. Black holes with a mountain-sized mass[15] of around 100 billion kilograms (10^{11} kg) should currently be on the verge of detonation.[16] Slightly larger and the mini black holes will continue emitting X-rays and gamma rays for many aeons to come. To date, none have ever been seen.

Whether or not mini black holes actually exist, there is absolutely no doubt that the principles of Hawking's theory are correct and that black holes do emit Hawking radiation. The combination of general relativity, quantum mechanics and thermodynamics mesh together so well that these ideas must play an important role in the fundamental make up of the universe. This remains the most important result linking quantum mechanics to gravity. It is among the most profound ideas in the history of physics.

A Singular Solution

What would it be like in the vicinity of a black hole? (We will consider an isolated black hole with no accretion disc, in order to avoid being roasted by radiation produced by material falling towards the black hole.) The first thing to realize is that black holes are not cosmic vacuum cleaners. Far from the event horizon of a black hole, its pull would be no different from that of a star of the same mass; an orbiting planet would follow an elliptical orbit, just like the planets in the solar system. Closer to the black hole, the planet's orbit would show significant precession.

If we were falling towards a distant black hole or in orbit around the black hole, we would feel nothing—we would be weightless. Stars have a surface that limits the depth of their gravitational well and once inside a massive body such as a star, the gravitational well starts to bottom out. But black holes are tiny by stellar standards, and they have no surface. As we get closer we may start to feel uncomfortable. In the Newtonian picture, the gravitational pull of the black hole is greatest on whatever part of our body is nearest to the black hole and this produces tidal forces that gradually increase as we approach. In Einstein's picture, in the vicinity of the black hole, space is stretched towards the black hole and simultaneously squeezed in the perpendicular directions.

In 1970, Arthur C Clarke published a two-page short story, *Neutron Tide*, constructed around the most outrageous play on words in science-fiction history. The tale is narrated by Captain Cummerbund who describes how the space cruiser *Flatbush* and three other starships were fighting the alien Mucoids when *Flatbush* strayed too close to a neutron star. The tidal gravity of the neutron star was so great that the space cruiser and its crew were ripped apart and all that the rescue mission recovered was a *star-mangled spanner*. This scenario is equally valid in the vicinity of a black hole. Just like Commander Karl van Rinderpest aboard the *Flatbush*, if we venture too close we will be torn apart before the vaporized remains of our body and our starship cross the event horizon and enter the abyss.

The collapsing core of a giant star might form a black hole whose event horizon is as small as 20 kilometres in diameter. But once the star is within its event horizon there is no escape; once inside there is no turning back. From here on, all roads lead to the centre of the black hole. Penrose's singularity theorems imply that any material body, such as a neutron star or even a hypothetical entity such as a quark star, must collapse if it finds itself within its own event horizon. The relentless squeeze of gravity will then inevitably crush the collapsing body out of existence. According to general relativity, the collapsed star and whatever subsequently falls into the black hole are compressed into a single point at its centre—the star's entire mass is located there in a point of infinite density where the warping of spacetime is so extreme that its curvature becomes infinite. Mathematicians refer to such a point as a singularity.

The mysteries within a black hole might be shrouded from the rest of the universe by the one-way barrier of the event horizon, but a point of infinite density, where the equations of general relativity break down, presents physicists with a serious problem.[17] Infinity is a familiar idea to mathematicians who routinely play with it in their mathematical proofs. Even so, it is necessarily a purely abstract concept that has no counterpart in the real world—when taking a physical measurement we cannot measure a value of infinity. Some quantities are extremely large, well beyond the reach of our instruments but, in reality, no quantity can be infinite. The physical existence of an infinite quantity would lead to logical paradoxes that would invalidate all the laws of physics. A singularity would be such an entity, a point where the forces of nature could not be understood, even in principle. If this were true, then anarchy would reign in a nonsensical Alice in Wonderland world down the gravitational rabbit hole.

Yet the singularities within black holes are sometimes discussed as though they do have a real physical existence. This is a mistaken understanding of what our theories tell us. General relativity models the universe very accurately but some care is required to interpret the meaning of the singularity at the centre of the black hole. The prediction of an infinite quantity must imply that the theory has been stretched beyond the domain where it gives a reliable representation of reality. In short, the singular point in the equations does not mean that there is a point of infinite density within a black hole, it implies that our best theory of gravity—general relativity—has been pushed beyond its breaking point.

General relativity seems to predict its own limits. There is good reason, however, to think that if general relativity were somehow combined with quantum mechanics to produce a quantum theory of gravity then the issue with singularities would be resolved. There are no stable atoms in classical physics. According to classical physics, the electrons orbiting an atomic nucleus would collapse into the nucleus in an instant. Quantum mechanics is required to explain the stability of atoms. Perhaps it is possible to stabilize the centre of a black hole in a similar way. Although we do not yet have a viable quantum theory of gravity, the need for such a theory was recognized long ago.

Spacetime Foam

Max Planck (1858–1947) laid the foundations of quantum theory in the year 1900. He had been struggling for several years to explain laboratory measurements of the colour and intensity of light that is emitted from an object as its temperature changes. In deriving a formula that explained the experimental results, he had been forced to conclude that the vibrations of the atoms in solids come in lumps or quanta. Planck knew at once that his revolutionary proposal would have seismic consequences. It would lead to the biggest upheaval in physics since the days of Isaac Newton. The cornerstone of quantum theory was a new fundamental constant of nature that relates the frequency of a vibration to the amount of energy that it carries. We know this constant as Planck's constant. It is always represented by the letter h and is a measure of the granularity of the universe. (For convenience, physicists often use the symbol $\hbar = \frac{h}{2\pi}$, which is known as the reduced Planck constant.) If Planck's constant were zero then the universe could be correctly described by the classical physics of Newton and Einstein. Although it is not zero it is very small, and this means that classical physics works well as a description of the macroscopic world of everyday life. When considering very short length scales, however—at the scale of atoms and below—quantum mechanics must be taken into account.

Planck realized that the existence of quantum theory would eventually demand a reconstruction of every branch of physics, including gravity. By combining his new constant with the other fundamental constants of nature, Newton's gravitational constant and the speed of light, Planck made a rough estimate of the distance scale on which quantum effects would play an important role in gravity. This distance is known as the Planck length. It has the symbol L_P and it is given by the following expressions:

$$(L_P)^2 = \frac{\hbar G}{c^3}, \quad L_P = \left(\frac{\hbar G}{c^3}\right)^{1/2},$$

where $\hbar$ is the reduced Planck's constant, also known as 'h bar', G is Newton's constant, the fundamental constant of gravity, and c is the speed of light in a vacuum, the fundamental constant of relativity. The Planck length is the distance scale on which quantum mechanics, relativity, and gravity are all simultaneously relevant. If we understood physics on this distance scale, then we might be able to calculate Newton's constant G, from some deeper, more fundamental physical principles, and we might also have an explanation for the weirdness of quantum mechanics. When we put in the numbers, we find that the Planck length is a mere 10^{-35} metres. If we scaled ourselves down to the size of a proton and simultaneously scaled a proton down by the same factor, the rescaled proton would still be much larger than the Planck length. This incredibly short distance is the fundamental unit of length in our universe.

Similarly, it is possible to work out the fundamental unit of time. It is known as the Planck time and it is the length of time that it would take for a particle travelling at the speed of light to travel a distance equal to the Planck length. The Planck time is 10^{-43} seconds. Quantum mechanics is a theory involving random fluctuations, and gravity is best described as spacetime curvature. So what do we get if we combine the two? No-one knows for sure, but some theorists imagine that space and time turn into a sort of frothy foam. The American physicist John Archibald Wheeler, who developed these ideas in the 1950s, described this quantum foam as:

> spacetime stirred into the writhing turbulence of myriad multiply connected domains.[18]

Of course, the reality could be quite different, and any genuine understanding of quantum gravity is yet to arrive. It is possible that space and time, as we understand them, emerge from some as yet unknown, deeper geometrical structures that are relevant on these minuscule scales.

Across the Event Horizon

Hawking derived a formula for the entropy of a black hole. It is expressed most succinctly in terms of the Planck length, the

Figure 7.12 Stephen Hawking

length scale on which quantum effects become important in gravity. If we cover the event horizon with an ultra-fine grid, such that it is divided into regions whose length and breadth equal the Planck length, how many such regions will there be? In other words, how many quantum granules of space does it take to cover the event horizon? The answer is

$$\frac{A}{\left(L_P \times L_P\right)^2},$$

where A is the surface area of the event horizon and $L_P \times L_P$ is the area of one square region of this mesh. Hawking deduced that the entropy of a black hole is one quarter of this number, that is a quarter of the area of the event horizon in units of the Planck length:

$$S = \frac{A}{4(L_P)^2}.$$

According to Planck's definition of the Planck length $(L_P)^2 = \frac{\hbar G}{c^3}$. So, substituting for $(L_P)^2$ gives the following expression for the entropy of a black hole:

$$S = \frac{c^3 A}{4\hbar G}.$$

Multiplying by Boltzmann's constant k gives the entropy in the appropriate units for Carnot's original formulation where entropy is expressed as a quantity of heat divided by temperature. This produces the expression

$$S = \frac{kc^3 A}{4\hbar G}$$

shown on the commemorative 50p piece in Figure 7.6. Hawking's formula implies that the entropy of a stellar mass black hole is truly astronomical, vastly greater than the entropy of all the material that collapsed and fell into the black hole.

When Stephen Hawking was diagnosed with motor neurone disease as a student in the 1960s, he was given just two years to live. Confounding these predictions, Hawking lived to become the world's most famous scientist since Einstein. Hawking (Figure 7.12) was appointed Lucasian Professor of Mathematics at Cambridge University in 1978. This was the post that had been held by Isaac Newton 300 years earlier. Hawking's illness was progressively worsening, and the university expected his appointment to be a short-term stop-gap measure. Confounding the sceptics yet again, Hawking retired in 2009 at the age of 67.

Hawking finally crossed the event horizon on 14 March 2018. His incredible determination to succeed in the face of any obstacle and his drive to live life to the full are an inspiration for us all. We should also remember his incredible communication skills and his lifelong commitment to those less able and less fortunate than himself. Hawking is buried in Westminster Abbey close to the tomb of Sir Isaac Newton.

8

Ripples in the Fabric of Things

Sometimes big things happen, and they echo. Those echoes crash across worlds. They are the ripples in the fabric of things. Often they manifest as storms. Reality is a fragile thing, after all.

NEIL GAIMAN, *World's End* (1994)

Black holes are black. Nothing comes out apart from Hawking radiation, which is so faint it is impossible to detect. We cannot see black holes. But what if we could feel the presence of these behemoths throwing their weight around in the depths of space? In 1916, Einstein had an idea that has enabled us to do just that.

Einstein's investigations that led to relativity were inspired by his desire for a better understanding of electromagnetism. His general theory of relativity describes gravity as the warping of space and time by massive objects. This might seem very different to electromagnetism but according to general relativity, gravity should exhibit many features that are analogous to electromagnetic effects. For instance, an accelerating electrically charged particle, such as an electron, generates ripples in the electromagnetic field that permeates space. When the ripples correspond to visible wavelengths we refer to them as light. Waves of other wavelengths are known as radio waves, microwaves, infra-red, ultra-violet, and so on. Einstein realized that gravity should produce similar effects. He reasoned that the acceleration of massive bodies must disturb the gravitational field which, according to general relativity, is spacetime itself. So vigorously whirling large masses around should generate ripples in the fabric of space that emanate from the gravitating system. Figure 8.1 illustrates how

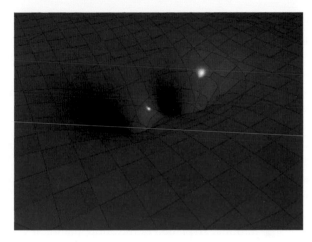

Figure 8.1 Schematic representation of the shape of space around two very massive compact objects in orbit around each other.

Figure 8.2 The passage of a gravitational wave. The positions of test particles change as a gravitational wave passes through space. As the gravitational wave passes, space is squeezed in one direction and stretched in the perpendicular direction.

these gravitational waves might arise; it shows a two-dimensional slice of space warped by the presence of two ultra-dense bodies, such as black holes. As the bodies orbit each other, space is distorted, and ripples of warped space stream outwards.

We live within three-dimensional space. A passing gravitational wave would have the effect of squeezing and stretching the space around us. Figure 8.2 is a sequence of eight snapshots showing how the positions of twenty-four test particles would be affected by a passing gravitational wave. The relative positions of the test particles change as a gravitational wave passes through space. The test particles are initially arranged in a circle. As the gravitational wave passes, space is squeezed in one direction and stretched in the perpendicular direction. The figure shows a single cycle of the wave. (The direction of the wave would be into or out of the page.)

We live in an environment where gravity is very important, and this gives a false impression of its strength. In fact, gravity is incredibly feeble compared to the other forces of nature, such as electromagnetism. It takes a planet-sized amount of matter pulling together for gravity to have an appreciable effect, and even then, it is easy to pick up metal objects with a small magnet, defying the gravitational attraction of the entire planet. If we place the forces of gravity and electromagnetism on a level playing field, we can see the true disparity in their strengths. The electrostatic attraction between an electron and a proton is around 10^{40} times the gravitational attraction between the two particles, which gives some idea of the astonishing weakness of gravity. It is only because gravity is always attractive, and therefore has a cumulative effect, that it becomes significant for sufficiently large concentrations of mass and thereby plays an important role in the large-scale structure of the universe.

The strength of the electromagnetic force means that it is easy to generate electromagnetic waves. When an alternating current is passed through an aerial, for instance, the electrons in the aerial, which is just a long thin piece of metal, move rapidly up and down. Accelerating the electric charges of the electrons in this way generates electromagnetic waves that we know as radio waves. This was first observed in the laboratory by Heinrich Hertz in 1887. We perform the same experiment every day when we turn on the radio or television or use our mobile phone. Generating gravitational waves is rather more difficult. The weakness of gravity means that an enormous mass must be given an almighty shaking to produce anything but the tiniest gravitational ripple.[1] Even Einstein was unsure whether it would ever be possible to detect them. The first indication that gravitational waves really do exist came from a star system far, far away.

The Laboratory at the End of the Universe

In 1974, astronomers Joseph Taylor (1941–) and Russell Hulse (1950–) searched systematically for pulsars with the giant Arecibo radio telescope in Puerto Rico and found many examples of these remarkable objects. Amidst the data was one whose behaviour seemed unusual. The pulses from pulsar beams are received with incredible regularity, but this one example was curiously different. The intervals between the pulses would increase for a while and then decrease again. There was a regular pattern to this behaviour, which would repeat every seven and three-quarter hours.

Hulse and Taylor had found a *binary* neutron star system. The system consists of two neutron stars in orbit around each other, and the pulsar belongs to one of them. As this neutron star approaches the Earth, the interval between the arrival of its pulses steadily decreases and then, as the neutron star recedes from Earth, the interval increases again. This is the well-known Doppler effect that we often hear with moving sirens. (No pulsar signals have been received from the other neutron star, so if it also generates a pulsar beam, it never points in our direction.)

The binary system was a great find because the pulsar's regularity has enabled astronomers to study the motion of the neutron stars and measure their properties with great precision. We know that the two collapsed stars have about the same mass, which is around one and a half times the mass of the Sun. They complete one orbit every 7.75 hours. This means that the orbit is quite small by cosmic standards. At their closest, the neutrons stars are separated by a distance that is slightly greater than the radius of the Sun, whilst at their greatest separation they are almost five times further apart. It is amazing to have such detailed information about the paths followed by objects that are 20,000 light years away, which is about a billion times further away than the Sun.

But the Hulse–Taylor neutron star system has much more to offer. The solar system is quite a sedate home. The planets serenely orbit the Sun, and Newton's theory of gravity explains their motion with great accuracy. Only in the case of Mercury was there a small hiccough that required a better theory than Newton's. As we have seen, the resolution was provided in 1915 by Einstein's theory of general relativity. Einstein's theory gives different predictions for the shape of planetary orbits, but the differences are tiny unless the planets are moving rapidly in a very strong gravitational field. As Mercury is the closest planet to the Sun, the relativistic corrections to its orbit are much bigger than for the other planets—but the effect is still very small.

The Hulse–Taylor system is a much more extreme gravitational environment than the solar system, so the relativistic effects are much greater. And the pulsar provides a built-in precision timekeeper. This gives astronomers a great physics laboratory in which they can put Einstein's theory to the test. When the paths of the two neutron stars were mapped out in detail, as expected their highly eccentric orbit was shown to be precessing, so it did not quite match the Newtonian prediction, but it agreed with Einstein's theory exactly. This was a great confirmation of our modern understanding of gravity and remains a significant triumph of Einstein's theory of general relativity. Even so, this effect had already been seen with Mercury many years earlier, so it was not quite headline news. But there was something else that was far more exciting.

The Hulse–Taylor binary neutron star system is a marvellous laboratory for astrophysicists. The orbital period of the two neutron stars can be measured with great accuracy. It is about seven and three-quarter hours, but it is decreasing by 76.5 microseconds every year. General relativity provides the explanation. As the two neutron stars whirl around each other at high speed, they generate gravitational waves, and the orbit of the neutron stars is shrinking due to the loss of the energy that is radiated into space in the form of these waves. When the rate at which the orbit would shrink due to the emission of gravitational waves was calculated, the predictions of general relativity matched the observations to perfection. This was the first confirmation of the existence of gravitational waves. Einstein's theory had triumphed again. In 1993, Hulse and Taylor were awarded the Nobel Prize in Physics for their discovery.

The neutron stars in the Hulse–Taylor system approach each other more closely every year as their orbit decreases in size due to the emission of gravitational waves. In 300 million years the two neutron stars are scheduled for their final encounter. Their meeting is sure to be an incredibly violent occasion.

A Blast from the Past

The Nuclear Test Ban Treaty was signed in 1963. To monitor compliance with the treaty, the United States launched the Vela series of satellites that were designed to search for gamma rays, which are the unmistakable signatures of nuclear explosions. The satellites soon began to detect occasional flashes of gamma rays, or gamma ray bursts, as they are known. These alarming and mysterious signals were immediately put under investigation.

By 1973, it was clear that the gamma rays were arriving from deep space, and the research was declassified by the military; astronomers have worked steadily to uncover their secrets ever since. As the gamma ray bursts typically last just a few seconds, this has proved quite a challenge. Success has depended on the rapid deployment of telescopes to study the lingering afterglow of an eruption following the detection of a burst of gamma rays by instruments aboard a satellite.

It is now clear that these bursts of gamma rays originate in some of the most violent cataclysms in the universe. These events are rare, but they are so powerful that we can detect them from the other side of the universe, thousands of millions of light years away. Gamma ray bursts have been divided into two categories. Most last for a few seconds and are known as long gamma ray bursts. The other category, the short gamma ray bursts, last for less than two seconds.

Last Tango in Deep Space

The long gamma ray bursts are believed to originate in the terminal collapse of a giant star and its transformation into a black hole. As the star collapses, it spins ever faster. When the black hole forms, it is just a few dozen kilometres across and spinning at almost the speed of light. Black holes are often portrayed as gaping like the jaws of Hell. But black holes can be very messy eaters. They are tiny by cosmic standards, so squeezing an entire star into one proves difficult, and some of the plasma shoots out at the poles, rather than entering the abyss; this material is compressed beyond nuclear densities and focused into two beams that are generated by the intense magnetic fields of the rapidly rotating plasma of the black hole's accretion disc. The beams spurt outwards at ultra-relativistic velocities, accompanied by an intense blast of gamma rays that blazes through the depths of space. A civilization on the other side of the universe that happens to be looking down the barrel of this gamma ray-gun may eventually pick up a brief trace of radiation signalling the death of a mighty star and the formation of a black hole.

Short gamma ray bursts have proved even more difficult to study, because they are so brief and much less powerful. On 3 June 2013, there was an important breakthrough. The gamma ray telescope aboard NASA's Swift satellite picked up a gamma ray burst that lasted just a tenth of a second. Nine days later, the Hubble Space Telescope took up the search for the origin of the gamma radiation. Hubble found a faint glow in a galaxy four thousand million light years distant, so the gamma ray burst had been produced in an event that occurred when the Earth was in its infancy.

Analysis of Hubble's images has shown that the short gamma ray burst was generated by a type of stellar explosion known as a kilonova, so named because they are around 1000 times as bright as a nova. Nevertheless they are just a hundredth to a tenth of the brilliance of a supernova. What sort of drama produces a kilonova, you might ask? They are generated by the merger of two neutron stars that have reached the climax of their cosmic dance. In 300 million years' time, this will be the ultimate fate of the binary neutron star system discovered by Hulse and Taylor. In addition to gamma rays, the collision of such hyper-dense bodies should generate gravitational ripples, and in the 1970s some physicists were wondering whether it might be possible to detect these ripples here on Earth.

Silent Whispers

Detecting electromagnetic waves is quite straightforward. We do it whenever we open our eyes, turn on the television, use Wifi, or heat a cup of tea in a microwave oven. But because gravity is so weak, detecting gravitational waves is a formidable technical challenge. The earliest attempts were made in the 1960s. Indeed, the American physicist Joseph Weber (1919–2000) announced in 1969 that he had constructed a device that was receiving gravitational wave signals on a daily basis from the central region of the Milky Way galaxy. His apparatus was built around an aluminium bar that would vibrate at its resonant frequency when a gravitational wave passed through it. There was a rush to build similar devices around the world, but unfortunately no other researchers could replicate Weber's results. Theorists were also quick to dismiss Weber's claims, questioning whether his equipment could possibly be sensitive enough to detect gravitational waves, and arguing that if such mighty cosmic fireworks were erupting on a regular basis throughout the galaxy, then surely, we would also see them through our telescopes. Nonetheless, Weber's false start generated world-wide interest in the possibility of detecting gravitational waves, whilst highlighting the perils of making premature announcements. Subsequent researchers would proceed with extreme caution.

Attention soon turned to the interferometer, the device invented by Michelson in the nineteenth century, as described in Chapter 5. Rainer Weiss (1932–) at MIT (Massachusetts Institute of Technology) was one of the champions of this approach. Weiss, who was an expert in the development of precision instruments, believed that large-scale modern interferometers might just have a chance of detecting gravitational waves despite the many technical hurdles that lay ahead. In his view, interferometers of sufficient sensitivity were feasible, but they would have to be several kilometres in length, so a price tag of hundreds of millions of dollars was inevitable.

Before funding on this scale could be secured it was vital to know what such an instrument might find. Could we be sure that it would find anything? It was true that detecting gravitational waves would confirm another prediction of general relativity, providing further support for our best theory of gravity. But gravity is so weak that only the most violent cosmic events such as black hole mergers would produce waves that might conceivably be detected. Could physicists be sure that black holes really exist? In the 1970s the evidence was not totally compelling. So what would such an instrument detect?

The theorist Kip Thorne (1940–) spearheaded the programme to improve our understanding of the violent cosmic processes that might generate gravitational waves. Thorne and his research teams based at Caltech (California Institute of Technology) examined possible sources, such as supernova explosions, neutron star collisions, and black hole mergers, so that physicists would have a better understanding of the signals that they were looking for and the feasibility of finding them.

Other physicists in Europe, including Ronald Drever (1931–2017), James Hough (1945–), and their team at the University of Glasgow in Scotland, and Heinz Billing (1914–2017), Karsten

Danzmann (1955–), and their team at the Max Planck Institute for Astrophysics in Garching, Germany, also took up the challenge. Their research into precision interferometers would lead to key technologies that would be incorporated into later machines.

Small prototype interferometers were constructed and step by step the technology advanced. During the 1980s, the NSF (National Science Foundation) in the United States funded feasibility studies to determine the practicality of developing a large-scale gravitational wave laboratory. Finally, in 1990, the NSF approved the construction of LIGO (Laser Interferometer Gravitational-wave Observatory) with a budget of USD 300 million and, in 1994, the particle physicist Barry Barish (1936–) was appointed as director of the project. Barish was chosen for his expertise in managing large-scale international particle physics projects and tasked with leading this fantastically ambitious venture to a successful conclusion.

Good Vibrations!

By 2002, two LIGO facilities had been constructed 3000 kilometres apart in the United States: one in the north-west at Hanford, Washington; the other in the south-east at Livingston, Louisiana. Two well-separated detectors are required to distinguish true gravitational wave signals from local background disturbances, so the two laboratories are vital components of a single enterprise. Travelling at the speed of light, a gravitational wave would take 10 milliseconds to cover the distance between the two sites, producing a similar signal at both, with a small time delay depending on the direction from which it arrived.

The LIGO detectors are L-shaped, consisting of two perpendicular 4-kilometre ultra-high vacuum pipelines. Figure 8.3 shows the LIGO facility in Livingston. One 4-kilometre arm recedes into the distance, while the other disappears into the trees to the right of the picture. Two mirrors within each arm are suspended

Figure 8.3 The LIGO facility in Livingston, Louisiana.

as test masses and the laser interferometer accurately monitors the distance between them. Space is alternately stretched and squeezed when a gravitational wave passes through the instrument, and this changes the distances between the mirrors. As one arm is stretched the other is squeezed, in the manner illustrated in Figure 8.2. The purpose of the interferometer is to measure these utterly minuscule changes.

The interferometer works as follows. A laser is directed at a beam-splitter that sends half the beam down each arm. Each half-beam travels 1600 kilometres in total, bouncing back and forth 400 times between the two mirrors in that arm, before the two half-beams are recombined. The interferometer is tuned so that the recombined half-beams completely cancel, with the peaks in the light waves of one beam meeting the troughs in the other, so no light passes to the photodetector. Whenever a gravitational wave ripples through the apparatus, however, the lengths of the arms alter very slightly, the distances travelled by the half-beams

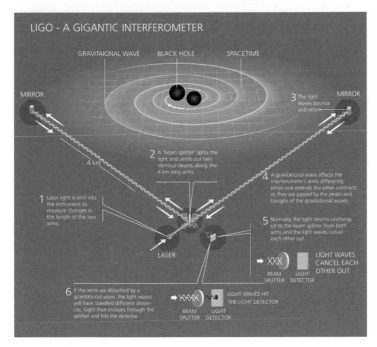

Figure 8.4 Schematic diagram of the LIGO gravitational wave detector.
Credit: © Johan Jarnestad/The Royal Swedish Academy of Sciences.

change and their relative phases shift. Now the cancellation is no longer perfect and some light seeps through to the photodetector, as illustrated in Figure 8.4.

LIGO is most sensitive to gravitational waves whose wavelength is comparable to the distance the laser beam travels within the interferometer. The peak sensitivity is for wavelengths between about 100 kilometres and 10,000 kilometres, corresponding to frequencies between 3000 and 30 hertz (cycles per second). We would expect such waves to be generated by huge masses travelling at relativistic velocities in gravitating systems of a comparable size to these wavelengths. Such systems include ultra-compact binaries formed of neutron stars or black holes as they spiral towards their final encounter.

During its first decade of operation, LIGO was not expected to see anything. It was a test bed that would enable physicists to understand their delicate instrument better and develop improved technologies that would further enhance its sensitivity. A major upgrade to Advanced LIGO that was scheduled for completion in 2015 would give physicists their first realistic chance of feeling the vibrations of space.

The overwhelming challenge faced by the LIGO team is to ensure that the tiny effect of a gravitational ripple is not swamped by the many possible sources of noise, both from within the machine and from its surroundings. Every feature of Advanced LIGO is designed to eliminate background noise and enable the faint gravitational wave signals to be extracted. The upgraded machine includes a higher-powered laser, and the test masses are 40-kilogram fused silica mirrors with a multi-layered optical coating that gives them almost perfect reflectivity while reducing their thermal noise. Each mirror is suspended by a sophisticated quadruple-pendulum system mounted on a seismic isolation platform that suppresses vibrations due to the tidal motion of the Earth's crust and actively cancels shifts due to micro-earthquakes. The sensitivity of Advanced LIGO is extraordinary, as it must be if there is to be any chance of success.

After decades of development, it was believed that Advanced LIGO would finally achieve the sensitivity required to detect gravitational waves. But could the theory be wrong? Are events such as black hole mergers so rare that you might have to wait a century or more before detecting one?

Echoes Crashing Across the Worlds

A rare, beautiful creature came sailing along the wave,
called from its throat to the land,
resounding loudly; the laughter was fearsome,
terrible on earth; her edges were sharp.

                                    The Exeter Book of Riddles, 33

On 14 September 2015, during the final calibration tests, and before the official start of its first observing run, a clear signal was detected by the interferometers in Hanford and Livingston. Researchers checked and re-checked the data, every possible source of a spurious signal was identified, and a statistical analysis of the noise within the machines suggested a false-alarm rate of just once in 200,000 years for such a strong signal.[2] This is how often a random spasm within the detectors would produce a comparable output. For several months rumours of a momentous discovery circulated in the scientific media, then finally on 11 February 2016 an official announcement was made—LIGO had received a signal. This first gravitational wave signal is designated GW150914, which simply encodes the date on which it arrived. Just over four centuries after Galileo first pointed his telescope at the stars, and in the centenary year of Einstein's general relativity, gravitational wave researchers had opened up a new window on the universe.

The signal was produced by the collision and merger of two distant black holes—violently whirling balls of warped space. The collision of these cosmic dervishes was like a stone being thrown into a still pond which generates a wave that races away from the point of impact. Ripples on a pond expand in ever increasing circles. Similarly, gravitational ripples expand through space as an ever-expanding spherical wave. This wave travelled an immense distance across the universe before it reached the LIGO detectors here on Earth. This was the first ever detection of a binary black hole system and, indeed, the most direct observation of black holes ever made.

Figure 8.5 shows a compilation of some of the LIGO data from this first signal.

Time is plotted along the horizontal axis in the graphs and is labelled in twentieths of a second. The signal was received in the early hours of the morning in the United States, but is time-stamped as 09:50:45 UTC, where UTC means *coordinated universal time*. This is essentially Greenwich Mean Time. The entire

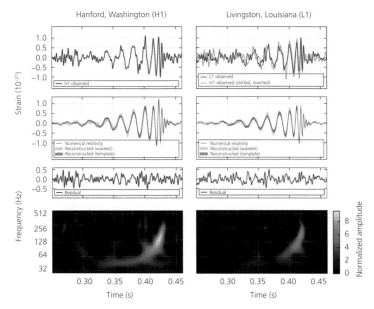

Figure 8.5 The first ever gravitational wave signal detected by the LIGO observatories[3].

signal lasted just one-fifth of the subsequent second, as shown in the graphs. Its brevity was the first indication that it originated in the terminal encounter of two very massive black holes.

The LIGO interferometer measures the change in distance between the mirrors in its arms due to a passing gravitational wave. The proportional change in distance is known as *strain*. In the top three rows of Figure 8.5 this strain is plotted on the vertical axis and is marked in units of 10^{-21}. The length of each arm is 4 kilometres, or 4×10^3 metres, so a single unit change in length on this scale represents a difference of $4 \times 10^3 \times 10^{-21}$ m $= 4 \times 10^{-18}$ m in the distance between the mirrors. The peak strain measured by the interferometer due to the first signal was about this size.

To gauge how minuscule this is, we can compare it to the size of a subatomic particle. The nucleus of an atom is composed of protons and neutrons, and the diameter of a proton is just under 2×10^{-15} m. So the Advanced LIGO interferometers are capable of measuring changes in distance that are less than one thousandth of the diameter of a proton.

The Livingston interferometer registered the wave 6.9 milliseconds before Hanford, which implies that it arrived from somewhere in the southern hemisphere. Figure 8.5 (top left) is the output from the Hanford detector. On the top right, this output is shifted to remove the time lag, then flipped and laid over the output from the Livingston detector. (The Hanford data are flipped to allow for the different orientation of the interferometers at the two sites.) The match is good, but not perfect. This is because of the unavoidable random noise within each detector.

The gravitational wave signal is disentangled from the noise in the graphs of the second and third rows. Physicists have generated a library of computer simulations composed of hundreds of thousands of templates that model black hole mergers and other violent cosmic scenarios. The second-row graphs show the waveform from the template simulation that gives the best match to the gravitational wave data for the two observatories. The waveform labelled *numerical relativity* is from a template giving a 99.9% fit to the gravitational wave data. The third-row graphs show the noise that has been subtracted out. (In other words, the graphs in the top row are the sum of the gravitational wave signal and the noise shown in the second and third rows.) Having identified the best fit to the data, it is now possible to deduce the physical characteristics of the black hole collision and merger from the parameters that were used to generate the matching template. Such comparisons reveal a great deal about the origin of the gravitational wave signals.

What Do the Signals Tell Us?

One of the beauties of gravitational waves is that they are un-affected by any matter they might encounter on their journey across the universe, so the signal we receive is unmodified since its origin. It therefore retains information about the cataclysm in which it was produced. It is simply a scaled down version of the original wave.

The signals from binary black hole mergers come in three phases: the *inspiral*, the *merger*, and the *ring-down*. The inspiral data encode many of the properties of the binary system, such as the masses of the convergent black holes. The data comprise a con-tinuous stream of gravitational waves that drain energy from the binary system causing the black holes to spiral together. The waves are generated by the warping of space as the black holes race around each other ever faster. Their frequency is twice the rate at which the black holes orbit, for essentially the same reason that the Moon's gravity raises two tides a day at each point on Earth.

So what do we know about the source of GW150914, the first ever signal? LIGO detected the final ten cycles of the inspiral cor-responding to the last five orbits prior to the merger of the black holes. The matching computer template tells us that the signal was produced by an encounter between black holes of 29 and 36 solar masses. Given this information, the initial amplitude of the gravitational wave can be calculated. By comparing this to the amplitude of the detected waves the distance to the merger event can be estimated. It turns out to be a staggering 1.3 billion light years.[4]

The frequency and amplitude of the gravitational wave signal rose dramatically during the final moments of inspiral, as shown in Figure 8.5 (second row). This information has been used to create the fourth-row graphs which plot frequency against time, with the amplitude of the waves colour-coded into the plots. The data show that the two black holes, which were both about 200

kilometres in diameter, orbited each other ever faster as they spiralled together. The gravitational wave signal approached four
hundred cycles per second as the black holes touched, and this
means they were racing around at almost half the speed of light
as they merged.[5]

During the *merger* phase, the newly merged black hole is rather
asymmetrical. It settles down with a final blast of gravitational
waves known as the *ring-down*, which leaves the black hole with a
perfectly spherical event horizon. The features of the merger and
ring-down can be used to determine the mass and spin of the re-
sultant black hole. This analysis shows that the two black holes
coalesced to form a rapidly spinning black hole of 62 solar masses.

Cosmic Spacequakes

So the computer models tell us that black holes of 29 and 36 solar
masses coalesced into a black hole of 62 solar masses. This might
look rather odd as 29 + 36 = 65, so three times the mass of the Sun
seems to have disappeared in the process. In fact, during the collision and merger some of the mass of the black holes was converted
into pure energy in the form of gravitational waves in a remarkable instance of Einstein's formula $E = mc^2$, which implies that
mass is just another form of energy. According to the computer
models, the mass of the final merged black hole is significantly less
than the total mass of the two colliding black holes when they
were well separated. Energy is conserved in all processes, so any
change in mass has to be taken into account when we balance the
books, which implies that the total energy carried by the expanding sphere of gravitational waves from this event is equivalent to
three times the mass of the Sun.

Still, you might protest: *if black holes are supposed to be black, how could
this mass escape from the black holes?*

It is perfectly true that a black hole is surrounded by its event
horizon and anything within the event horizon is on a one-way

trip into the abyss. Once inside nothing comes out again. The solution is that the gravitational waves did not come out of the black holes, they formed outside. The gravitational waves carry away the binding energy that is released as the black holes approach and merge.

Gravitational binding energy is released whenever two bodies fall together under their mutual gravitational attraction. We can see this in action if we hold a stone aloft and then allow it to fall under gravity. Initially, the stone has a certain amount of potential energy. This is the amount of binding energy that will be released if the stone is allowed to undergo free fall. When the stone is dropped it accelerates towards the ground and the potential energy is transformed into kinetic energy. When the stone hits the ground, its kinetic energy is converted into other forms of energy, such as sound waves, and is eventually dispersed as heat. The ultimate source of this heat is the binding energy between the stone and the Earth. To raise the stone back to its initial height we would have to input exactly the same amount of energy as the binding energy that was released by its fall.

Figure 8.6 The Chelyabinsk meteor.

Even in the vicinity of the Earth we can see dramatic examples of the release of gravitational binding energy. When a meteor falls to Earth it produces a blazing trail as it burns up in the Earth's atmosphere. The most spectacular example in recent times was the meteor that raced across the skies of the Chelyabinsk region of Russia in the Urals on 15 February 2013. It was captured on dashboard cameras as it blazed across the early morning sky. Figure 8.6 shows the moment when the space rock exploded creating a brilliant flash and a shockwave that shattered windows throughout the region. The ultimate origin of the meteor's energy was gravitational binding energy.

Perhaps, the clearest analogy to the colliding black holes comes from a totally different area of physics—the formation of a hydrogen atom from a proton and an electron. There is an electrostatic attraction between a positively charged proton and a negatively charged electron, and when they bind to form a hydrogen atom one or more photons are emitted. These electromagnetic waves carry away the binding energy of the atom. According to Einstein's formula, this energy is equivalent to a small amount of mass, which means that the combined mass of a well separated proton and an electron is slightly greater than the mass of the hydrogen atom they form when bound together. The energy of the photons equals the mass that is lost when the hydrogen atom forms.[6] This corresponds to just over one millionth of one percent of the mass of the atom. (To separate the proton and the electron the same amount of energy must be put back in, which is possible in various ways.[7])

Now back to the colliding black holes. There is a very strong gravitational attraction between them. The closer they approach each other the greater the amount of binding energy that is released, and this energy is radiated away in the form of gravitational waves. Consequently the mass of the bound pair of black holes is less than the total mass of the two black holes when they were well separated. Although the distance and energy scales are vastly different, there is a clear parallel between the binary black hole and the hydrogen atom. The hydrogen atom is bound by the electromagnetic force and the binding energy of the proton and electron is carried away as electromagnetic waves. The binary black hole system is bound by gravity and the binding energy is carried away as gravitational waves. One big difference is that the merged black holes cannot be separated again.

It is truly astonishing that GW150914, the first signal detected by LIGO, was generated in just a fraction of a second by a black hole merger in which three times the mass of the Sun was converted into pure energy in the form of gravitational waves. We can use Einstein's formula $E = mc^2$ to see just how much energy this is. Here E represents energy in joules (J), m is mass in kilograms, and c is the speed of light in a vacuum in metres per second. As the Sun's mass is 2×10^{30} kilograms, the total mass converted into gravitational waves was around 6×10^{30} kilograms. To find the energy, we multiply this figure by the square of the speed of light in a vacuum, which is 300,000 kilometres, or 3×10^8 metres, per second. So the energy carried by the gravitational wave is

$$E = mc^2 = \left(6 \times 10^{30}\right) \times \left(3 \times 10^8\right) \times \left(3 \times 10^8\right) \text{ J}$$
$$= 54 \times 10^{46}\text{J} = 5.4 \times 10^{47}\text{J}.$$

We can also work out the power output of the merging black holes. This is the rate at which this energy conversion took place. It can be calculated by dividing the total energy of the gravitational waves by the time period in seconds during which it was generated. Figure 8.5 shows that most of the signal arrived within a twentieth of a second. Multiplying the gravitational wave energy by twenty, we reach a figure in the region of 10^{49} watts. At its peak, the power output of gravitational waves generated by the merging black holes may have been somewhat greater than this.

Such colossal figures are hard to comprehend. For comparison we can consider the luminosity of all the stars in the visible universe. The luminosity of the Sun is 4×10^{26} watts. There are a few hundred billion stars in the galaxy, but the Sun is brighter than most. The luminosity of the galaxy is roughly 10^{11} times that of the Sun, or about 4×10^{37} watts. There are an estimated 10^{11} galaxies in the visible universe, which gives the total luminosity of all the stars in the visible universe as something like 4×10^{48} watts. So, incredibly, for a fraction of a second, the energy emitted as gravitational waves by this one black hole merger was significantly greater than the combined luminosity of all the stars in the visible universe.

Gravitational waves carry vast amounts of energy distributed over enormous regions of space. Anyone who happened to find themselves in the vicinity of the black hole merger would have been torn apart by the stretching and squeezing of the gravitational waves. When the gravitational waves arrived on Earth 1.3 billion years later however, their stentorian roar had expanded into a sphere of 1.3 billion light years radius and dwindled to the faintest of whispers. These ripples are so small that feeling the quake of such mighty encounters in the furthermost depths of space ranks as one of the most astonishing technical accomplishments in human history. In recognition of this amazing achievement Rainer Weiss, Kip Thorne, and Barry Barish were awarded the 2017 Nobel Prize in Physics for the design and construction of LIGO and the detection of gravitational waves.[8]

Although Weiss, Thorne, and Barish were the pivotal figures who pushed for the development of LIGO and steered this monumental project to a successful outcome, LIGO is a truly international project involving thousands of researchers around the world and is a great example of modern science at its best.

This was just the start of the age of gravitational astronomy. Much more was to come.

The Primaeval Thunderclap

Bababadalgharaghtakamminarronnkonnbronntonnerronntuonnthunn
trovarrhounawnskawntoohoohoordenenthurnuk!

James Joyce, *Finnegans Wake*

The epoch-making gravitational wave signal GW150914 was the first of three signals detected by LIGO during its first observing run which ended on 19 January 2016. LIGO resumed operation for its second run at the end of that November and bagged another five signals over the course of the next few months. All eight emanated from colliding pairs of black holes. Then, on 17 August 2017 the two LIGO detectors, and a newly commissioned Franco-Italian detector called VIRGO, constructed near Pisa in Italy, heard something quite different. This primaeval thunderclap, designated GW170817, was much longer than anything heard before, rumbling on for over a minute. The small time difference in the signals received by the three detectors enabled physicists to determine the direction towards its source. Moreover, just 1.7 seconds after the arrival of the gravitational wave signal, NASA's Fermi Gamma-ray Space Telescope detected a short gamma ray burst from the same region of sky. Telescopes around the world were marshalled to locate the glowing embers of the cataclysm and within twelve hours the source was located. It was a kilonova produced by the collision of two neutron stars (Figure 8.7).

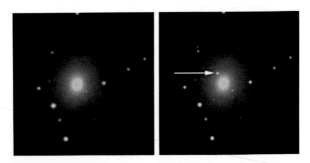

Figure 8.7 Left: Overexposed image of the galaxy NGC 4993 before the explosion. Right: The arrow indicates the kilonova.

Like a Joycean thunderclap, the rumbling gravitational wave signal heralded the dawn of a new age.[9] This was the first time that an optical counterpart for a gravitational wave signal had been detected, and it became one of the most observed objects in the history of astronomy. The age of multi-messenger astronomy had begun.[10] The gravitational waves and the short gamma ray burst were produced by the collision and merger of a pair of neutron stars 138 million light years away in a galaxy known as NGC 4993. For millions of years the neutron stars had gradually spiralled together, losing energy through the emission of gravitational waves, just like the binary system discovered by Hulse and Taylor. LIGO and VIRGO detected these waves as they increased in amplitude and frequency during the final 100 seconds before their crunching encounter. Each of the neutron stars had about a twentieth of the mass of the black holes whose collision produced the GW150914 signal. Because they were so much less massive, they spiralled together more slowly, and this is why the signal was detected for so much longer. Nonetheless, their combined mass is believed to be too great for them to exist as a single neutron star, so they are presumed to have collapsed to form a black hole after their thunderous collision (Figure 8.8).

Figure 8.8 Artist's impression of a neutron star collision such as the one that produced GW170817.

The nuclei of most atoms are forged in nuclear reactions within stars and dispersed in supernova explosions. Such events were long thought to be the source of almost all the elements in the periodic table. Recently, however, doubts have arisen about whether it is possible to create the heaviest elements in this way. According to computer simulations, many elements, such as silver, iodine, platinum, gold, and uranium, are far more abundant than can be explained by supernovae. So where do these atoms come from? It is now believed that many of the heavier elements that we find so useful formed in neutron star collisions that occurred billions of years ago prior to the formation of the solar system. Although these spectacular collisions are infrequent, they provide the extreme conditions necessary for the synthesis of these nuclei and the explosive power to disperse them throughout the galaxy in vast quantities. The observation of further examples of these collisions will help to determine whether this picture is correct.

The Rhythms of the Cosmos

How many black holes are there in the Milky Way galaxy? A very rough estimate of the number of burnt-out giant stars that have undergone the ultimate terminal collapse over the long history of our galaxy suggests a figure of at least one hundred million. Yet, only about twenty or so have been identified. A half-century ago, the X-ray emissions of Cygnus X-1 were interpreted as due to the blazing hot plasma of a black hole's accretion disc. Since then just a couple of dozen other black holes have been discovered in our galaxy, and the information that we have gleaned about them is rather indirect.

Gravitational wave signals have already added significantly to our knowledge of black holes. These data offer a much sharper picture of these weird objects, providing precise details about their mass and spin. Black hole physics is now an observational science.

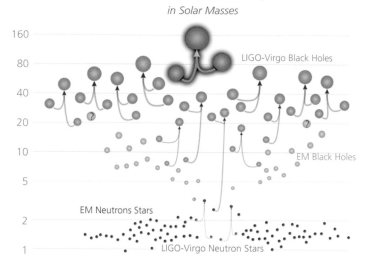

Figure 8.9 The chart plots the known masses of neutron stars and black holes. Those known via electromagnetic radiation, such as pulsar signals or X-ray emissions, are labelled EM Neutron Stars and EM Black Holes. Those known through gravitational wave signals are labelled LIGO-Virgo Neutron Stars and Black Holes.

Although black hole mergers are rare events, gravitational wave signals can be detected from vast distances. All the signals seen so far have arrived from distant galaxies, very remote from our own. Figure 8.9 is a chart of the masses of the known neutron stars and black holes including those whose demise has been witnessed by LIGO and VIRGO.

LIGO has examined the theory of general relativity like never before. We already know that Einstein's theory matches the workings of the physical world better than Newton's. But earlier comparisons were all made in rather gentle gravitational systems where the differences between the two theories are small. By contrast, the information gleaned from gravitational waves begins

with the most extraordinary prediction of general relativity—
the existence of black holes—and tests the theory in the ultimate gravitational scenarios—black hole mergers. Can we still
rely on the theory to explain what happens in these extreme
circumstances? The computer simulations that are used to interpret the LIGO data are based on general relativity and the fact
that the templates match the wave data shows just how wonderful the theory is. So far, general relativity has passed every
test.

Other areas of astrophysics are also being challenged. The signal GW190521 was the mightiest black hole merger seen so far;
black holes of 66 and 85 solar masses merged to create a 142 solar mass whopper. These beasts are shown at the top of Figure 8.9.
Current models of stellar evolution cannot account for such massive black holes. The models suggest that very massive stars are
so unstable that they completely blast themselves apart in their
terminal supernova explosions, leaving no black hole remnant
whatsoever. Prior to the gravitational wave data, astrophysicists
had believed that black holes of between 50 and 150 solar masses
could not form from collapsing stars. Their existence is yet to be
explained.

Gravitational waves also offer the best prospects for investigating ultra-dense matter. We do not know the maximum mass that
can be supported as a neutron star, because their internal structure and composition is not yet fully understood, and the core
density of a neutron star is beyond anything that can be probed in
the laboratory. But multi-messenger observations of neutron star
collisions will provide clues that could reveal how matter behaves
under such extreme conditions.

In August 2019 the LIGO and VIRGO detectors picked up the
signal GW190814, produced by a collision and merger of a 23 solar
mass black hole with an ultra-compact object of 2.6 solar masses.
But what was the lower mass object? Was it a neutron star, a black
hole, or something else? (It is marked with a question mark in

Figure 8.9.) This object had more mass than any known neutron star, but less than any known black hole. It was in the twilight zone that has been referred to as a *mass gap*. There is no reason, however, to assume that compact objects do not fall within this mass range. Rather, it is the identity of such objects that remains obscure. When the gravitational wave observatories find further examples, they will no doubt help to solve the mystery.

The early discoveries from gravitational wave observatories are already enhancing our knowledge of neutron stars and black holes. What else can we expect to see? Prior to the successful completion of LIGO, estimates of how often black hole or neutron star mergers might occur within range of the detectors were almost total guesswork. Finding signals from supernova explosions seemed a safer bet. It was at least known that supernovae happen. A supernova explosion in our own galaxy is an infrequent occurrence, but supernovae are regularly seen in other galaxies. The Franco-Italian detector VIRGO was named with a nod to the Virgo cluster of galaxies. This is the nearest large cluster, 50 million light years away. Ironically, LIGO and VIRGO are yet to lose their virginity to a supernova explosion. But it is only a matter of time before a star blows its top, close enough for its rippling spacequake to be felt here on Earth. The Virgo cluster contains about 2000 galaxies, so it will probably be home to the first supernova that produces a sufficient gravitational blast wave to be detected. This will provide valuable information about the processes deep within the core of an exploding star and could dramatically improve our models of the final stages of stellar evolution. (Incidentally, a perfectly spherical collapse will not produce any gravitational waves, so supernova spacequakes will only be produced by supernovae that are somewhat lopsided.)

We should certainly expect the unexpected, and this may be the most exciting prospect. Every new window on the universe has opened our eyes to new exotic wonders of the cosmos. We have visited some of the greatest of these discoveries. Radio telescopes revealed neutron stars and their pulsar beams, gamma ray detectors spotted kilonova blasts produced by colliding neutron stars, and X-ray telescopes gave us the first evidence for black holes such as Cygnus X-1. Who knows what gravitational wave observatories might find?

Standard Candles and Standard Sirens

The strength of an optical signal, whether it is received by a cell in our retina or a pixel in the CCD chip of a camera, depends on the amount of energy deposited by the light wave. The same is just as true for any other region of the electromagnetic spectrum, such as radio waves. Detecting gravitational waves is rather different. Interferometers measure changes in distance as a gravitational wave passes by, so they are responding to the size, or amplitude, of a gravitational wave, not the amount of energy it deposits. This has a certain advantage when improving their performance.

Astronomers refer to celestial objects of a fixed luminosity as *standard candles*, a name coined by Henrietta Swan Leavitt, as we will see in Chapter 9. They are important because they provide astronomers with a measuring rod to the stars. Doubling the distance to a standard candle will reduce its luminosity to one quarter of its previous brightness. This is because as the distance to the standard candle increases its light spreads over a sphere of increasing radius. The sphere's surface area increases as the square of its radius, so the intensity of the light falls off as the *inverse* square of the distance to the standard candle. This must be so if energy is to be conserved. Consequently, to see a standard candle at twice the distance, we must increase the sensitivity of our optical instruments by a factor of four.

The gravitational wave equivalent of a standard candle is known as a standard siren. The amplitude of a gravitational wave falls off as the inverse of the distance to its source not the inverse square, (this is because the energy carried by a wave is proportional to the square of its amplitude), which means that if the distance to a standard siren were doubled, then the size of the gravitational wave signal would be reduced by half. Consequently, to detect a standard siren at twice the distance, we only need to increase the sensitivity of our gravitational wave detector by a factor of two.

Puzzle 10 LIGO currently detects a gravitational wave signal once a week. If its sensitivity were increased by a factor ten, how many signals could it expect to detect each week?

A New Wave of Detectors

It is early days for gravitational wave astronomy. LIGO averaged less than one signal per month during its first two observation runs; three in the first four-month run and eight in the second nine-month run. The sensitivity of the detectors is being incrementally improved and each enhancement expands the volume of space that is accessible. Even small improvements bring large dividends. During LIGO's third observing run, which began on 1 April 2019, LIGO was averaging more than one signal each week.

In March 2020, LIGO's third observing run was suspended due to the corona-virus pandemic. In the next run, the LIGO and Virgo detectors will be joined by KAGRA (Kamioka Gravitational Wave Detector), a new detector that has been constructed in the Kamioka mine in Japan. As further observatories come on line around the world, a more precise determination of the source of gravitational wave signals will be possible, increasing the opportunities for multi-messenger observations. KAGRA is the first gravitational wave observatory to be built underground to reduce environmental disturbances, and whereas LIGO and VIRGO operate at room temperature, KAGRA will be cooled to just 20 degrees above absolute zero which will reduce the random thermal noise in the mirrors and other components of the machine. Other detectors are in the pipeline, such as LIGO-India, also known as INDIGO, which will be constructed in the state of Maharashtra in central India.

The most obvious way to improve the sensitivity of gravitational wave detectors is to increase the size of the interferometer; bigger generally means more expensive, however. There is a European plan to construct a third-generation gravitational wave observatory known as the Einstein Telescope[11] (Figure 8.10) that will be ten times as sensitive as Advanced LIGO. The generation tag is somewhat arbitrary as the development of gravitational wave interferometers has been a long process of gradual improvement. But the early prototypes are regarded as the first-generation detectors, the current LIGO and VIRGO detectors are considered as the second generation, and the third-generation machines will

Figure 8.10 Artist's impression of the Einstein Telescope.

be designed from scratch incorporating all the tried and test technology developed in the earlier machines.

Answer to Puzzle 10 Improving LIGO's sensitivity by a factor ten would mean the distance to the faintest signal it could discern would be increased by a factor of ten. This would increase the volume of space that was accessible by a factor of ten cubed, or 1000. Assuming that sources of detectable signals, such as black hole mergers, are evenly distributed throughout the universe, the detection rate would increase by the same factor to around 1000 per week.

The Einstein Telescope will consist of three 10-kilometre interferometers arranged in an equilateral triangle. The interferometers will be housed in underground tunnels to reduce seismic noise and tidal gravity effects due to the movement of nearby masses and cooled to cryogenic temperatures to reduce the thermal noise in the mirrors. Increasing the sensitivity by a factor of ten increases the accessible volume of space by a factor of one thousand; this should increase the detection rate from the current average of one event per week to 1000 events per week. The source of the first gravitational wave event detected by LIGO was 1.3 billion light years distant. It is 13.8 billion years since the Big Bang. So a tenfold increase in sensitivity would produce a similar signal from an event 13 billion light years away, at the very dawn of the universe. The Einstein Telescope would be capable of detecting black hole mergers from the entire visible universe. Two sites are currently under consideration as a home for the new device; one in Sardinia, the other in the Meuse-Rhine region close to the borders of the Netherlands, Belgium, and Germany. Construction is expected to begin in 2026 leading to the first observations in 2035.

Meanwhile, researchers in the United States are planning the construction of an even grander gravitational wave observatory that has been named Cosmic Explorer.[12] It will consist of an inter-ferometer with 40-kilometre arms in an L-shaped configuration, similar to LIGO.

In the next chapter we will travel to the ends of the universe and the beginning of time, visiting some of the weirdest beasts in the entire cosmos.

9

Across the Universe

The chessboard is the world; the pieces are the phenomena of the universe; the rules of the game are what we call the laws of Nature.

THOMAS HENRY HUXLEY, *Lay Sermons, Addresses, and Reviews* (1870) 'A Liberal Education'

Einstein realized as soon as he had completed general relativity that his new theory of gravity would offer a tool to investigate the structure and evolution of the universe as a whole.[1] Einstein's starting point was his assumption that the universe is essentially the same everywhere when considered on the very longest distance scales. This assumption allowed Einstein to treat the matter within the universe as though it were a uniform fluid. Although observational evidence to support this idea was not available at the time, it has proved to be a good approximation to the very large-scale structure of the early universe.

When Einstein investigated the consequences of the model, his calculations suggested that the universe must be either expanding or contracting. But this jarred with his intuitions. Einstein believed that the universe must be eternal and static, so he introduced a new term into his equations—the cosmological term—representing a new hypothetical force that would balance gravity over extremely long cosmological distances. He later described this as the biggest blunder of his life. Others such as the Russian theorist Alexander Friedmann (1888–1925) and the Belgian Georges Lemaître (1894–1966) were happier to follow where the equations of general relativity led. Lemaître, in particular,

Figure 9.1 Albert Einstein and Georges Lemaître.

was keen to promote the idea that the universe sprang into existence from an incredibly dense state—the *primordial atom*, as he called it—billions of years ago and has been expanding ever since (Figure 9.1).

Perhaps it is no surprise that Einstein believed the universe to be static. When he completed his masterpiece in 1915, astronomers knew far less than they do today. Since then, the world has changed in many ways, but the universe has been transformed beyond all recognition.

From dark sites a lustrous shimmering path can be seen arching across the night sky. This is the Milky Way. In the constellation Andromeda there is a faint smudge of light. One hundred years ago, it was called the Great Nebula of Andromeda. The distance to this starry cloud was completely unknown so, along with other misty patches in the night sky, it was presumed to be a region of gas and stars within the Milky Way. Indeed, astronomers believed that the Milky Way formed the universe in its entirety. This would soon prove to be wildly inaccurate. Our understanding of the true scale of the cosmos began with the work of a relatively unknown astronomer Henrietta Swan Leavitt (1868–1921).

Figure 9.2 The Harvard Observatory Computers.

The Scale of the Cosmos

Leavitt was a computer, which is not to say that she was some sort of early robot. The mechanical devices we know today as computers are named after a now obsolete occupation. Computers were people, mainly women, who were employed to perform mundane repetitive calculations. Over the course of a couple of decades in the late nineteenth and early twentieth centuries Edward Charles Pickering (1846–1919) recruited about eighty women to work as computers for astronomy projects at the Harvard College Observatory (Figure 9.2).

Leavitt completed her Harvard degree course in June 1892 and was presented with a certificate to say that, if she had been a man, she would have qualified for a BA degree (Figure 9.3). Leavitt was almost deaf following a childhood illness, but she was financially secure, so she offered to work for Pickering for free. He was delighted to take her on, and she became one of his computers. She was later paid about USD10 a week.

Figure 9.3 Henrietta Swan Leavitt.

Today this all sounds rather exploitative and demeaning, but Pickering was actually quite progressive in encouraging women to be involved in such work. The project he gave to Leavitt was to analyse the stars in the Magellanic Clouds as recorded on a series of photographic plates taken at Harvard's Boyden Observatory in Arequipa in Peru. The Large and Small Magellanic Clouds are now known to be dwarf galaxies that orbit our own Milky Way galaxy (Figure 9.4). They are visible from the southern hemisphere and were unknown to Europeans before Magellan's circumnavigation of the globe in the sixteenth century.

Figure 9.4 Panoramic view of the Milky Way above the European Southern Observatory (ESO) site in Chile. The Magellanic Clouds can be seen on the far left of the image.

Many stars vary in brightness over time, sometimes in a regular way and sometimes quite unpredictably. Leavitt's task was to study these variable stars and, in particular, a type known as Cepheid variables. The luminosity of these stars varies regularly over a period of several days. You may have seen the closest Cepheid variable. It is the Pole Star or North Star, known to astronomers as Polaris (Figure 9.5). Polaris lies at a distance of 430 light years and varies in brightness over a period of about four days. Cepheid variables are named after another northern star Delta Cephei, which is 880 light years away and varies in brightness over a period of five days and nine hours.

When comparing stars, we cannot easily tell whether a star is very bright but at an immense distance, or not so bright and relatively nearby. For instance, Sirius is the brightest star in the night sky, much brighter than Polaris, but we know today that Sirius is one of our closest stellar neighbours, just over eight and a half light years distant, whereas Polaris is fifty times further away. If these two stars were equally distant then Polaris would easily be the brighter of the two.

Figure 9.5 How to find Polaris.

When Leavitt was working, over a century ago, the distance to the Small Magellanic Cloud was unknown. But she assumed that as it consists of a dense cluster of stars, these stars must all lie at essentially the same distance from us. This was key to her great discovery because it meant that all the Cepheid variables in her sample were equally distant, so any differences in their brightness would reflect true differences in their luminosity. Leavitt spotted a pattern. She found that the brighter a Cepheid variable the longer the period over which its brightness varies. Her discovery is known as the period-luminosity relationship for Cepheid variables. Leavitt's results were announced by Pickering in 1908 in a Harvard College Observatory circular and this was followed up with a second paper in 1912.

Leavitt's discovery of a pattern in how the brightness of an obscure type of star varies might not sound particularly exciting. But she could see just how important it would be. Leavitt referred to Cepheid variables as *standard candles*. She realized that they could solve the problem of determining distances in the universe. It is easy to monitor the brightness of a Cepheid variable and determine its period of variation. Once this is known, Leavitt's period–luminosity relationship can be used to deduce the intrinsic brightness of the star. We know how luminosity decreases with distance—it falls off as the inverse square of distance—so by measuring how bright the star appears in the sky we can determine how far away it is. Furthermore, Cepheid variables are very bright stars, so with powerful telescopes they can be observed at great distances.

Leavitt's work would indeed lead to a revolution in our understanding of the distance scales of the universe. But tragically she died of stomach cancer in 1921 at the age of 53 just too soon to see the revolution come about. In 1924 Edwin Hubble (1889–1953) used her relationship to show that our Milky Way galaxy is just one among many galaxies. Using the Hooker telescope at the Mount Wilson Observatory, California, then the world's largest telescope, Hubble found Cepheid variables in the Great Nebula of Andromeda. His estimates of their distance demonstrated that this nebula is, in fact, a separate galaxy, much more distant than any stars in our own galaxy. He then showed that other similar nebulae are also independent galaxies. Hubble's discovery dramatically enlarged the universe overnight.[2]

Leavitt's period–luminosity relationship still forms an important rung in the cosmic distance ladder, which is the basis for measuring the size of the cosmos. We now know that our galaxy is around 100,000 light years across, and we live about 26,000 light

Figure 9.6 The Andromeda galaxy, which is thought to look similar to our own Milky Way galaxy.

years from its centre. The Andromeda galaxy, our closest galactic neighbour apart from a few dwarf galaxies, is about 2.5 million light years distant and is similar in size and shape to our own galaxy (Figure 9.6).

Henrietta Swan Leavitt never obtained a doctorate or an academic position. Yet, if she had lived a few more years, she almost certainly would have been nominated for a Nobel Prize.

Hubble used Leavitt's standard candles to gauge the distance to numerous galaxies. In the photograph shown in Figure 9.7 Hubble is pointing to a Cepheid variable in the outskirts of the Andromeda galaxy. By the end of the decade Hubble had made another cosmos-shattering discovery. He found that the characteristic lines in the spectrum of light from distant galaxies are shifted to the red end of the spectrum which indicates that the galaxies are racing away from us. Hubble showed that the greater the distance to a galaxy, the faster it is receding, so the entire universe is expanding, just as Lemaître and the equations of general relativity had predicted.[3]

Figure 9.7 Edwin P. Hubble Papers, Huntington Library, San Marino, California.

In 1990, NASA launched its orbiting space telescope, and it was named the Hubble Space Telescope in honour of Edwin Hubble. The telescope has given us unprecedented views of the heavens. The Hubble Ultra-Deep Field (Figure 9.8) is one of its iconic images. This extremely long exposure photograph is the result of pointing the telescope at an apparently blank region of space for a total of twenty-three days over a ten-year period. In one small patch of sky, with less than 1% of the Full Moon's area,[4] it reveals 10,000 objects, most of which are extremely distant galaxies. This is a tiny sample of the estimated 100 billion galaxies in the visible universe.

To the Ends of the Universe

In the early years of radio astronomy, it was difficult to locate the precise source of any radio emissions that were being received. In 1963, an ingenious technique was used to pin down the exact

Figure 9.8 Hubble Ultra Deep Field.

position of one strong radio source catalogued as 3C 273, by timing the moment when its signal disappeared as it passed behind the Moon. The Dutch astronomer Maarten Schmidt (1929–) used the data from these occultations to track down the visible counterpart of the radio source with what was then the world's largest optical telescope at the Hale Observatory on Mount Palomar in California.[5] Schmidt then recorded the object's visible spectrum to determine its chemical composition. To his great surprise, the spectral lines did not match any known elements; they were completely mysterious.

Eventually, Schmidt realized that the lines were the usual lines produced by elements such as hydrogen and helium, but they had undergone an enormous red shift. The wavelength of each line was 16% greater than those measured in the laboratory. This huge red shift implied that 3C 273 was an incredible 2 billion light years distant, making it the most luminous object ever seen at that time.

3C 273 was the first example of a new class of astronomical objects that appear point-like, just as stars do, but with enormous red shifts. They were named *quasars*, which is an abbreviation of *quasi-stellar radio sources*. If they really were located at the vast distances that their red shifts suggested, then their energy output had to be gargantuan. Even some of the most daring astrophysicists of the day, such as Fred Hoyle (1915–2001), questioned the interpretation of the red shift–distance relationship in these cases, as the consequences appeared to be so outlandish.

The controversy persisted through the 1960s. It was not until the advent of much better telescopes, such as the Hubble Space Telescope in the 1990s, that astronomers could identify the galaxies that surround the quasars, proving that they really are located within galaxies at immense distances. We now know that quasars are the outward visible signs of incredibly violent activity at the centres of extremely distant and therefore faint galaxies. Furthermore, quasars fluctuate in brightness over very short periods of time—sometimes as short as a few hours—which means that the central powerhouse of the quasar must be very small indeed; just a few light hours across. They are the most extreme examples of what have become known as active galactic nuclei.

There are much closer active galactic nuclei that offer clues to the nature of the beasts that lie within; some appear to be the source of enormous jets of material spewing into intergalactic space. Cygnus A is one of the most powerful radio sources in the sky. It is 600 million light years away—a huge distance, but still much closer than the quasars. The radio waves that it emits come from two vast lobes of plasma that are ballooning outwards as two oppositely directed jets plough into intergalactic space (Figure 9.9). The object producing the jets is clearly exceptionally stable, as the jets have been fired in the same straight lines for

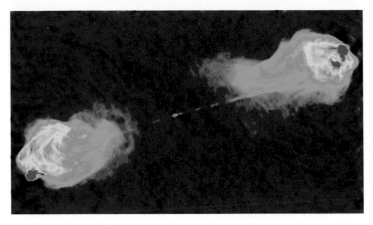

Figure 9.9 False-colour radio image of the hyper-luminous radio galaxy Cygnus A, showing intense jets and lobes that are thought to be issuing from a supermassive black hole.

millions of years. (The galaxy from whose heart the jets emanate is not visible in the image.)

The Ultimate Power Source

Quasars are the brightest objects in the universe. Their energy output is utterly staggering; some are thousands of times more luminous than the entire Milky Way galaxy or over 25 trillion times as luminous as the Sun. This luminosity is equivalent to transforming several stars' worth of material into pure energy in the form of electromagnetic radiation every year in accordance with Einstein's relationship between mass and energy. And all this from an object that appears to be a few light hours across—about the size of the solar system. So what could possibly be powering these colossal beacons from the edge of the universe?

The sheer scale of the power output of quasars rules out most possible sources of energy. In any process that releases energy, an equivalent amount of mass must be lost, as we have seen. In familiar processes, such as chemical reactions, the mass loss is tiny. Even in an explosive reaction such as the ignition of dynamite (TNT), the energy released accounts for just a few millionths of 1% of the mass of the TNT.

In nuclear reactions the conversion of mass into energy is far more substantial. Nonetheless, just 0.1% of the mass that is locked up in a uranium nucleus is released as energy in the fission reactions in the core of a nuclear power station. The Sun's energy is produced by the nuclear fusion of hydrogen into helium, which converts 0.7% of the mass of the hydrogen into energy. This reaction releases more energy than any other nuclear fusion process, but it is insufficient to explain the luminosity of quasars. There is another possibility, however—gravitational binding energy. The British astrophysicist Donald Lynden-Bell (1935–2018) saw this as the solution to the enigma posed by quasars and other active galactic nuclei.[6] In 1969, Lynden-Bell suggested that the stupendous energy output of a quasar could only be explained by stars or gas clouds falling into a rapidly spinning black hole located within the innermost core of a galaxy. No other object could conceivably release such a vast quantity of energy from such a small region of space. As material falls towards a black hole it is stretched and crushed in the intense gravitational field and a large proportion of its mass is released as energy before its final plummet. The exact amount depends crucially on how fast the black hole is spinning. The faster the black hole spins, the deeper the debris will sink into its gravitational well before reaching the event horizon—and the closer this final approach, the greater the amount of binding energy released. In the most extreme case of a black hole whirling at the maximum possible rate, up to 42% of the mass of the infalling plasma is released as energy before it disappears over the event horizon. And we can certainly expect that black holes will be spinning extremely fast.[7]

A statistical analysis of the population size of quasars in the distant universe suggests that they blaze away for millions, or even hundreds of millions, of years. If they have swallowed several stars' worth of material every year for this length of time, then their accumulated mass must be huge. Lynden-Bell tied all the evidence together and proposed that what we see as a quasar is the outward sign of a supermassive black hole of, perhaps, 100 million solar masses, as it rends and devours the stars and gas clouds in its neighbourhood.[8] This black hole would be surrounded by a vast accretion disc, where the shredded stars race around in a cosmic Catherine wheel of epic proportions. At its centre there would be more fireworks. The intense magnetic fields in the accretion disc act like cosmic particle accelerators, focusing and accelerating two intense jets of plasma almost to light speed as they blast outwards from the poles of the black hole. Spinning like a supermassive gyroscope, the black hole's axis would remain stable over the aeons, which accounts for the straightness of the jets in the Cygnus A system shown in Figure 9.9.

Quasars are all extremely distant, billions of light years away and, as we peer out into space we are looking back in time, so these blazing beacons are a feature of the distant past; they only exist in the early universe. Lynden-Bell suggested that, if his proposal were correct, there should still be supermassive black holes in the nearby galaxies of today. Having gorged themselves for billions of years on the material in their vicinity, these monsters might now be napping quietly in their lair, but it should still be possible to find the tell-tale footprints of these mighty beasts.

We are now sure that Lynden-Bell was correct. What we see as a quasar or other active galactic nucleus is the radiation emitted from the centre of a distant galaxy where a supermassive black hole is devouring the surrounding stars and gas clouds. The Chandra X-ray satellite has provided further evidence that all galaxies do indeed harbour a supermassive black hole. The accretion disc of a black hole is so hot that it emits X-rays, and this is the unmistakable signature of black holes, as Salpeter and Zeldovich predicted; only in the most extreme environments is matter heated to the enormous temperatures required for X-ray emission. Figure 9.10 shows the Chandra Deep Field South X-ray

Figure 9.10   Chandra Deep Field South X-ray image.

image that was released in 2017. It was produced by the Chandra telescope staring for a total of 7 million seconds at a small region of sky. It is the X-ray equivalent of the Hubble Ultra Deep Sky image. Every dot and smudge represent one X-ray source and most are believed to be accretion discs of supermassive black holes, each one residing within the heart of a galaxy billions of light years away.

It seems likely that all galaxies, including our own Milky Way, went through a rowdy, exuberant quasar phase in their early history, before settling into a quieter existence in later life when their central black hole has consumed most of the material in its immediate neighbourhood. To clinch the case for supermassive black holes, we should look at the evidence for one that is much closer to home. Is there really a supermassive black hole at the centre of our own galaxy?

Flying Teapot

Have a cup of tea, have another one, have a cup of tea
High in the sky, what do you see?
Come down to Earth, a cup of tea
Flying saucer, flying teacup
From outer space, Flying Teapot

DAEVID ALLEN, *Flying Teapot*

The hundreds of billions of stars in our galaxy form a disc with several spiral arms, but the greatest concentration of stars is found towards the centre. The galaxy is home to stars at every stage in their life. There are vast dust clouds and gaseous nebulae where new stars are born, clusters of bright young stars, aging stars surrounded by planetary nebulae, as well as bloated red giants and vast numbers of stellar relics, such as white dwarfs and neutron stars.

Figure 9.11 shows a beautiful region of the night sky in the constellation Sagittarius. Part of the constellation forms an asterism known to amateur astronomers as the *Teapot*. The steam from the spout of this flying teapot is formed of the many nebulae and gas clouds in the direction towards the centre of the galaxy. The exact centre of the galaxy is indicated by a cross in the figure, just near the spout. Radio astronomers have named this region Sgr A*, an abbreviation that singles it out as the most powerful source of radio signals (A) in the constellation of Sagittarius (Sgr), with the star (*) added to emphasize its special nature. In our galaxy, this is where the action is. Our immediate cosmic neighbourhood is incredibly quiet; the Sun is surrounded by oceans of empty space—it is over four light years to the nearest star. By contrast, within one light year of the centre of the galaxy there are, perhaps, a million stars. These include many burnt-out stellar remnants such as neutron stars and black holes, as well as many luminous blue supergiants.

There are occasional bursts of X-rays from SgrA*, which gives us a hint of what is lurking there. But the galactic centre is shrouded in hot gas, which blocks the visible light emitted from

Figure 9.11 The night sky in the direction towards the centre of the galaxy, including the *Teapot* asterism in the constellation of Sagittarius.

the stars in this region of the galaxy. Fortunately, infra-red radiation is much better at penetrating the murk. Figure 9.12 shows a false-colour image taken with NASA's SOFIA infra-red telescope spanning 600 light years around the galactic centre. The bright white patch just right of centre is the extremely hot gas and dust around SgrA*. The German astronomer Reinhard Genzel (1952–) is an expert in infra-red astronomy and has pioneered the study of this region of the night sky. In the early 1990s, Genzel used the European Southern Observatory's 3.5-metre New Technology Telescope in Chile to zoom into the innermost heart of the galaxy. He found that the stars right at the centre, in the vicinity of SgrA*, are racing around at incredible speeds. If the distribution of matter in the galaxy were uniform, we would expect the stars near the centre to be travelling much more slowly,[9] but the closer in that we peer, the faster they seem to be moving. Genzel's data showed that a huge amount of mass is concentrated at the exact centre of the galaxy, and its location appears to be fixed, while all else whirls frantically around.

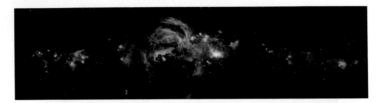

Figure 9.12 The galactic centre.

Genzel's observations were followed up by the American astronomer Andrea Ghez (1965–) at UCLA (University of California, Los Angeles). She has used the world's largest telescopes—the
two 10-metre Keck telescopes in Hawaii—to track the positions of
these stars for over twenty-five years and it is now clear they are
orbiting a hidden giant. Ghez has pioneered the use of state-of-
the-art techniques such as adaptive optics to produce ultra-high-
resolution images; adaptive optics improves the resolution of the
images from the Keck telescopes by a factor of about twenty.

The stars right at the centre of the galaxy are moving so quickly
that, over the course of just a few years, it has been possible to plot
out significant segments of their orbital paths. The closest neighbours to Sgr A* are racing around at up to 5 million kilometres
per hour.[10] As well as tracking their motion across the sky, their
velocity towards or away from us can be determined from the
Doppler shift of their light. This has enabled Ghez and her team
to calculate accurate trajectories of these stars in three dimensions. By imaging the stars every few months since 1995, they have
calculated the orbits of a dozen or so of the stars closest to the
centre; their remarkable results are plotted in Figure 9.13. One
such star, designated SO-2, completes its highly eccentric orbit
in just over sixteen years, and it has now been monitored over
the course of an entire orbit.[11] In May 2018, SO-2 reached the
innermost point of its orbit, passing within 18 billion kilometres,
or seventeen light hours, of the central object, which is about four
times the distance between Neptune and the Sun. At this point,

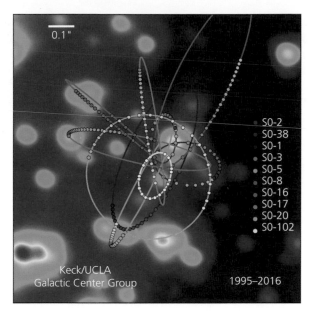

Figure 9.13 The stars around the galactic centre.

the star was travelling at 7,650 kilometres per second, or 2.5% of the speed of light.

The time taken by SO-2 and these other stars to complete each lap of their orbit is determined by the size of the orbit and the mass of the beast they are orbiting. This relationship is encapsulated by Kepler's Third Law which enables us to work out the mass of the hidden body. The best estimate from the orbital data suggests that the stars are orbiting a monster of 4.1 million solar masses. There is only one viable conclusion. The dust clouds are hiding a supermassive black hole. No other body of such enormous bulk could possibly reside within the orbit of SO-2.

This was certainly enough to convince the Nobel Prize Awards Committee. In 2020, half the Nobel Prize in Physics was shared between Andrea Ghez and Reinhard Genzel for providing conclusive proof that a supermassive black hole resides at the centre

of our galaxy. The other half went to Roger Penrose for his
lifelong work on the theory of black holes and, in particular,
his proof of fifty years earlier that black holes are an inevitable
consequence of general relativity.

Fierce Phlegethon and the Pit of Hell

> *Of four infernal rivers that disgorge*
> *Into the burning Lake their baleful streams;*
> *Abhorred Styx the flood of deadly hate,*
> *Sad Acheron of sorrow, black and deep;*
> *Cocytus, nam'd of lamentation loud*
> *Heard on the rueful stream; fierce Phlegethon*
> *Whose waves of torrent fire inflame with rage.*
>
> JOHN MILTON, *Paradise Lost*, BOOK II

We can watch the stars racing around a supermassive black
hole. But what if we could actually see the black hole itself? This
was the ultimate challenge that Shep Doeleman (1967–) of MIT
(Massachusetts Institute of Technology) set himself. Imaging the
event horizon of a supermassive black hole would require an instrument of unparalleled resolution with at least 5000 times the
resolving power of the Hubble Space Telescope.[12] Doelman has
coordinated an incredibly ambitious international effort to piece
together just such an instrument. It is known as the *Event Horizon Telescope* (EHT). The EHT collects data in the far infra-red or
microwave region of the spectrum. It combines the simultaneous synchronized observations by radio telescopes at eight sites
across the globe (Figure 9.14) acting together like one giant re-
ceiver almost as large as a hemisphere of the Earth. Writing in
2009, Doeleman claimed that, using this Earth-sized network of
radio telescopes, it is 'almost certain that direct imaging of black
holes can be achieved within the next decade'.[13]

The black hole at the centre of our galaxy is certainly big. But it
is a mere baby by cosmic standards. It is believed to be spinning relatively slowly—less than 10% of the maximum possible rate—so

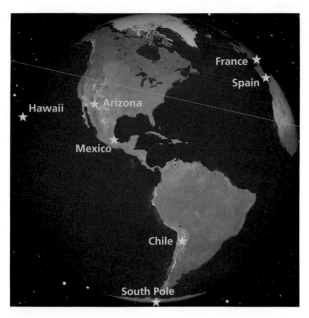

Figure 9.14 Sites of radio telescopes forming the Event Horizon Telescope.

its event horizon will extend almost to its Schwarzschild radius. As black hole masses are usually quoted in solar masses, it is easy to calculate their size as a multiple of the Sun's Schwarzschild radius which is 3 kilometres. The Sgr A* black hole is 4 million solar masses, so its Schwarzschild radius is 12 million kilometres. By comparison, the actual radius of the Sun is 700,000 kilometres.

At the centre of the Andromeda galaxy there is a supermassive black hole whose mass is estimated as a whopping 200 million solar masses, giving it a Schwarzschild radius of 600 million kilometres. But Doeleman set his sights on a supermassive monster on an altogether grander scale. It lies within the core of a relatively close active galaxy, the nearby supergiant elliptical galaxy known as M87 which lies at the heart of the Virgo cluster of galaxies 50 million light years away.

Figure 9.15 M87 and its jet.

M87 has grown to enormous proportions through the merger of numerous galaxies over its 13 billion-year history. It is notable for a huge jet of plasma that projects 5000 light years from the core of the galaxy and is blasting towards us at close to the speed of light (Figure 9.15). There is a jet pointing in the opposite direction, but this is very much fainter. The material spewed outwards by the jets over vast aeons of time has formed lobes that now stretch 250,000 light years into intergalactic space. Like the jets of Cygnus A, they emanate from the centre of the galaxy and are believed to originate in the rapidly spinning plasma of the supermassive black hole's accretion disc. The jets shoot out along the spin axis of the black hole, which is inclined just 17° away from pointing directly towards us.

On 10 April 2019 Doeleman kept his promise of a decade earlier when the EHT published its first image (Figure 9.16). It shows

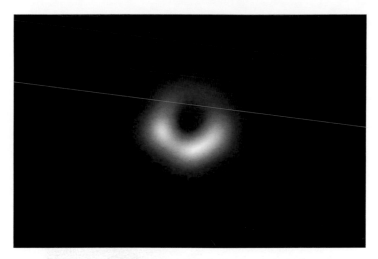

Figure 9.16 The M87 supermassive black hole as revealed by the Event Horizon Telescope. Microwave data collected by the EHT has been used to generate this computer-enhanced false-colour image.

the mighty supermassive black hole at the centre of M87 looking like the savage and inescapable pit of Tartarus enveloped by Phlegethon, the seething river of fire.

The orange ring in the image is the accretion disc of superheated plasma blazing away at a temperature of 6 billion degrees. This ring of plasma is swirling around the huge black hole. Even at close to the speed of light it takes several days for the plasma to complete a single circuit. Because the plasma is moving at relativistic speeds, the material in the ring that is travelling towards us, at the bottom of the image, is much brighter than the material that is travelling away from us at the top of the image. (It is for the same reason that the M87 jet that is moving towards us at close to the speed of light looks much brighter than the opposite jet that is moving away from us.) In the centre of the ring we can see a sphere. This is the event horizon of the black hole. Strictly speaking, it is somewhat larger than the event horizon and is referred to as the shadow of the event horizon. This

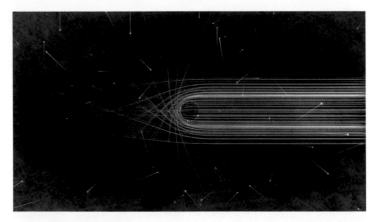

Figure 9.17 Artist's impression of how the gravitational bending of light in the vicinity of a black hole produces the event horizon shadow. The image shows the photon paths that would reach a viewer located to the right of the black hole.

is because the image is produced by light rays whose paths have been gravitationally bent around the black hole, as depicted in Figure 9.17.

Astronomers have revised their estimate of the mass of the M87 supermassive black hole using the results produced by the EHT. We now know that M87 is home to a black hole of titanic proportions whose mass is 6.5 billion times that of the Sun. Its diameter of 40 billion kilometres is over four times the size of Neptune's orbit around the sun, and over 125 times the size of Earth's orbit, so this huge black hole is comparable in size to the entire solar system.

Churning the Fabric of Spacetime

We are travellers on a cosmic journey, stardust, swirling and dancing in the eddies and whirlpools of infinity.

PAUL COELHO, *The Alchemist*

The further you look the weirder it gets. The existence of supermassive black holes of astonishing proportions is now an accepted feature of the universe. Every galaxy seems to have one of these amazing objects in its core. Over vast aeons of time galaxies merge and grow, which suggests that some galaxies should be home to more than one supermassive monster. The quasar OJ 287, 3.5 billion light years distant, is thought to be an example of just such a system.[14] OJ 287 is believed to consist of a gargantuan 18 billion solar mass black hole that is orbited by a second black hole with less than one hundredth of this mass—a mere 150 million solar masses. The primary black hole is surrounded by a large accretion disc which the other black hole hurtles through twice during each twelve-year orbit. This produces a plume of plasma and a blast of radiation over a trillion times the Sun's luminosity that has been detected periodically here on Earth for over a century.

The flares do not have the simple periodicity that would be produced by an elliptical Keplerian orbit; their timing matches a wildly non-Newtonian orbit deduced from a model based on general relativity, as shown in Figure 9.18.[15] This orbit has been used to successfully predict when the secondary black hole will pass through the accretion disc to produce brilliant outbursts from the system, such as in December 2015 and July 2019. At their closest approach, which occurred in July 2019, the two black holes are separated by just nine times the Schwarzschild radius of the primary or 3,250 times the distance between the Earth and Sun.[16] The highly eccentric orbit exhibits key characteristics of general relativity; it shows enormous precession, with the axis of the ellipse shifting by 39° each orbit. Furthermore, the giant whirling primary black hole drags space around like the monstrous whirlpool Charybdis and, as the smaller black hole—the fearsome Scylla—careers around the giant maelstrom, the plane of its orbit sweeps around in the churning spacetime. This is due to frame-dragging or the Lense–Thirring effect that was barely discernable by Gravity Probe B in Earth orbit, as described in Chapter 6, but it is dramatic in this extreme binary system. Finnish physicist Mauri Valtonen (1945–) and his colleagues have used the data from the flares to deduce the scale of the frame-dragging,

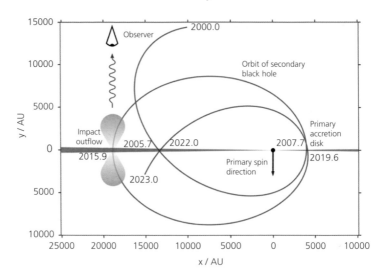

Figure 9.18 Diagram of the OJ 287 system showing the secondary black hole's orbit. We see the system from the direction perpendicular to the plane of the primary's accretion disc which is shown as the red horizontal line. Distance scales are in astronomical units (AU)—a distance equal to that between the Earth and Sun.

and through its effect on the orbit of the secondary black hole they have determined that the primary spins at one-third of the maximum rate for a black hole with its immense mass. The precise timing of future outbursts will provide more opportunities to refine our understanding of the orbit as well as further tests for general relativity.

The binary black hole orbit is gradually shrinking due to the emission of gravitational waves. (The energy lost in these emissions is 10,000 times that lost due to the secondary black hole blasting through the accretion disc.) In 10,000 years the two monsters Scylla and Charybdis will finally merge, sending an almighty rumble of gravitational waves out through the universe. This is just an instant from now in cosmological terms, but rather a long time for us to wait. There are, however, 100 billion galaxies in the visible universe. If each undergoes just one such epic encounter during its long history, their gravitational wave signals should be passing our way several times each year. What will it take to feel these tremendous vibrations?

Supersize Me!

To detect the gravitational ripples from such super-size mergers we need instruments of a comparable size to the systems in which they are generated. So the European Space Agency (ESA) is planning to put a gravitational wave observatory in space; LISA (Laser Interferometer Space Antenna) will consist of three spacecraft orbiting the Sun in a triangular formation, a mother and two daughter craft each separated by 2.5 million kilometres of empty space. They will form a precision interferometer with lasers monitoring the exact distances between the test masses floating within the mother and daughter craft. Each passing gravitational wave will change these distances very slightly and this will be detected by the interferometer.

The LISA Pathfinder mission (Figure 9.19) was launched in December 2015 as a stepping stone towards LISA. It was devised to test the technology and demonstrate the feasibility of constructing an interferometer in space. LISA Pathfinder released two cubic test masses to float freely within the spacecraft and used its laser interferometer to measure their separation, monitoring their positions to within one hundredth of a nanometre.

In the Pathfinder mission the test masses were located just 40 centimetres apart, whereas the three LISA craft will be separated by millions of kilometres. But the interferometer will measure their separation just as accurately, so its sensitivity will scale up in proportion to its increased size. ESA announced in June 2016 that the technology trialled by LISA Pathfinder had performed beyond expectations, which means it will certainly be capable of detecting gravitational waves when deployed by the LISA mission.

Figure 9.19 The technology package on the LISA Pathfinder satellite, with the interferometer monitoring the distance between two free-floating test masses composed of a gold-platinum alloy.

LISA is scheduled for launch in 2034 as part of ESA's Cosmic Vision programme.

LISA will greatly enhance our ability to study gravitational waves. It will detect signals invisible to LIGO and other ground-based gravitational wave detectors, as it will be sensitive to ripples with much longer wavelengths that are produced by much larger systems. LISA will pick up signals from tightly bound binary systems formed of compact objects, which may be white dwarfs, neutron stars, or black holes. These signals will be detectable when there is still an appreciable separation between the compact bodies, weeks or even months before their terminal collision and merger, which will greatly aid the visual identification of the merger event, opening up new possibilities for multi-messenger astronomy. LISA will allow the binary system's position in the sky to be determined and the time of merger to be accurately predicted, so that telescopes will be ready and waiting for the firework display. Watching such collisions as they happen will provide further clues about the structure of these weird ultra-compact bodies. As an important bonus, LISA will add to our knowledge of fundamental physics by providing stringent new tests for general relativity.

The most exciting prospect offered by LISA is the possibility of feeling the angry rumblings of the titans of the cosmos, the supermassive black holes. LISA will detect these beasts ripping up and swallowing the stars in their vicinity. Most marvellous of all, it will detect the awesome mergers of supermassive black holes in systems similar to OJ 287. LISA will be capable of detecting supermassive black hole mergers from the entire visible universe. No-one knows how frequently such signals will arrive.

The creation and early growth of supermassive black holes remains rather mysterious. There is believed to be an upper limit on the mass of a star of around 200 solar masses, and therefore an upper limit on the mass of a black hole that is produced through stellar collapse. A black hole must grow with the passage of time by accumulating material from the space around it. But there is a huge gulf between the hundreds of solar masses at the top end of stellar collapse models, and the millions and billions of solar masses seen in supermassive black holes. How could these black holes grow so enormous in the limited time available since the dawn of the universe? LISA will certainly help to answer such questions. LISA will feel the birth pangs and growth spurts of supermassive black holes, providing clues to how they formed and their relationship to quasars in the early universe. We can look forward to finding out much more about these monsters of deep space. LISA will also take us back to the very early universe and help to improve models of the immediate aftermath of the Big Bang.

The Big Bang Theory

The universe might be expanding, but the idea that it emerged from an incredibly hot ultra-dense state billions of years ago was once rather controversial. The theory was nicknamed the Big Bang by one of its vocal critics, the astrophysicist Fred Hoyle. It is a catchy name, which is why it has stuck, but it is also rather misleading. The Big Bang is sometimes represented as an explosion within the universe, and this is definitely *not* the case. The universe did not begin as a pre-existing container holding a cosmic egg from which the material that forms the stars and galaxies burst forth. Such a picture leads to the misconception that the Big Bang happened at a particular place. In fact, every point in space is equally distant from the Big Bang, and this distance is 13.8 billion light years. The theory actually claims that the universe in its entirety—space, time, and matter—began at the Big Bang.

It helps to consider the analogy of a balloon that is being blown up (Figure 9.20). The biggest difference is that the surface of a balloon is two-dimensional, whereas space is three-dimensional. As the balloon expands, every point on its surface moves away from every other point and, the further apart two points are, the faster they recede from each other—just like the galaxies in the real universe. We can run the expansion backwards, until every point on the balloon coalesces into a single point, which represents the origin of this rubbery universe. From this perspective, we can see that every point on the balloon universe is equally distant from its origin and, in fact, that the balloon Big Bang happened everywhere simultaneously.

There is remarkably good evidence for the Big Bang. First, as Hubble discovered, the universe is expanding. It is also worth noting that the universe does not appear to contain any objects, such as stars, that are older than 13.8 billion years. This is an important consistency check on the theory.

Figure 9.20 Balloon model of the expanding universe.

But there is also very good independent observational evidence. The universe would have been much hotter and denser in the past. For the first couple of minutes, it was a nuclear furnace in which fusion reactions created deuterium (heavy hydrogen), helium, lithium, and traces of other very light elements. These conditions did not last long enough for heavier atoms to be cooked up; all the heavier elements were generated much later in stars, supernova explosions, and neutron star collisions. It is possible to measure the amounts of deuterium and helium and other elements in the universe, and the observations closely match expectations from modelling the Big Bang.

A Noise Annoys!

Arno Penzias (1933–) and Robert Wilson (1936–) discovered the most compelling evidence for the Big Bang in 1964 while employed by Bell Labs in New Jersey (Figure 9.21). They were working on a sensitive horn antenna that was designed to receive the faint radio signals from early satellites, but it was plagued by a constant background noise that they were unable to eradicate.

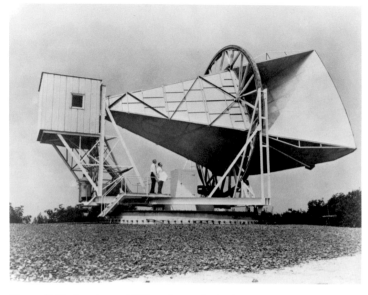

Figure 9.21 Penzias and Wilson beneath the horn antenna at Holmdel, New Jersey.

After discounting all terrestrial sources of interference, they showed that the signal was not arriving from the Sun, or even a source elsewhere within the galaxy, as it was uniform across the entire sky. They concluded it must originate in the furthest depths of space. They had discovered the cosmic microwave background, the relic radiation left over from the era immediately following the Big Bang. The microwave background had been predicted in the late 1940s by astrophysicists Ralph Alpher (1921–2007) and Robert Herman (1914–1997), although Penzias and Wilson were unaware of this.

So where do all these microwaves come from? After the era of nuclear synthesis, the expanding universe consisted of a plasma soup of charged particles, composed largely of hydrogen and helium nuclei and free electrons. This soup would have been awash with photons—the fundamental particles from which light is formed—bouncing around, continually scattering off the nuclei and electrons. Any electron that bonded with a nucleus to form an atom was immediately kicked out of the atom by a passing photon.

After 380,000 years of expansion, the matter cooled to around 3000 degrees—cool enough for hydrogen atoms to form. The universe was now transparent. The photons no longer had sufficient energy to interact with the matter, leaving them with the statistical pattern of energy and wavelengths characteristic of radiation emitted by matter at 3000 degrees. This radiation has raced across the universe for billions of years, during which time the universe has expanded in size by over 1000 times, and the radiation has undergone an enormous red shift due to this expansion. What set out as visible light at the furthermost extremities of the universe is now detected as radiation with a wavelength over 1000 times longer, in the microwave range; it is now identical to the electromagnetic radiation emitted by matter with a temperature less than 3 degrees above absolute zero—less than one thousandth of the temperature when it last interacted with matter.

The microwave background has an almost perfectly uniform temperature of 2.726 K across the whole sky. But there are tiny variations in the temperature, and these have been mapped by a series of space probes, the NASA satellites COBE and WMAP, and more recently the Planck satellite launched by ESA in 2009. These probes have looked back in time to an era immediately after the Big Bang to see the universe long before the formation of stars. Just as the light from stars and galaxies allows us to see how they were distributed when their light was emitted, so the microwave background gives us a picture of the distribution of matter during the epoch when it was emitted. The Planck satellite has plotted very small temperature variations over the entire sky, as shown in Figure 9.22; the red regions are 0.00002 degrees warmer than average and the blue regions are 0.00002 degrees cooler than average. The fact that the microwave background is almost perfectly uniform implies that matter was just as uniformly distributed

Figure 9.22 The tiny variations in the cosmic microwave background mapped out across the sky by Planck.

during the early universe.[17] The slight temperature variations correspond to slight variations in the density just 380,000 years after the Big Bang. This non-uniformity, or anisotropy, tells us how clumpy the matter was at this time; the slightly denser regions are the seeds that would grow into clusters of galaxies as the universe evolved.

The microwave background encodes vital information about the structure of the cosmos. It tells us the age of the universe, and this is how we know 13.8 billion years have elapsed since the Big Bang. The microwave background also provides information about some of the deepest unresolved mysteries of existence.

Most of the Universe Is Missing!

When we look up into a clear night sky, we see myriads of stars. Most are probably orbited by a family of planets, but the mass of a planet is tiny compared to the mass of a star. For instance, the Sun is over 1000 times the mass of Jupiter, the largest planet in the solar system, and all the other planets would easily fit within Jupiter. It might seem safe to assume that most of the mass of the universe is contained within its stars. Yet, this is far from the case.

Astronomers believe much of the mass of the universe exists in some form that is invisible. For this reason it is known as *dark matter*.

The first suggestion that much of the mass of the universe does not emit light dates back to the maverick astronomer Fritz Zwicky (1898–1974) in 1933. Overwhelming evidence has accumulated in the years since then. In the 1970s Vera Rubin (1928–2016) measured the rate of rotation of spiral galaxies (Figure 9.23). Her results consistently showed that galaxies rotate so fast they would fly apart if they were not held together by the gravitational attraction of much more matter than can be seen in the form of stars.

More recently gravitational lensing—the bending of light from distant galaxies, discussed in Chapter 6—has been used to estimate the mass of galaxy clusters and again the mass comes out much greater than would be expected from the light we can see. But the most accurate determination of the mass of the universe derives from the analysis of the cosmic microwave background.

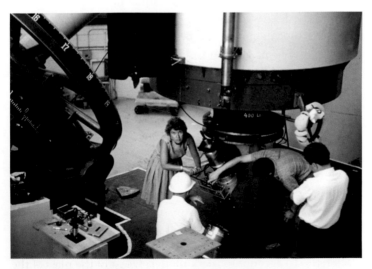

Figure 9.23 Vera Rubin in 1965, while working at the Lowell Observatory in Flagstaff, Arizona.

It is perhaps not so surprising that the universe contains matter that is invisible to us. But it certainly is surprising that there is far more dark matter than ordinary matter. According to data from the Planck satellite, dark matter makes up 84% of the mass of the universe. It seems that the galaxies that we see are just islands in an ocean of dark matter, or perhaps isolated mountain peaks above a layer of clouds.

What is It Made of?

We have a very good understanding of the fundamental physics of the universe. Almost all its features can be explained by two incredibly successful theories. Einstein's general relativity is still our best theory of gravity. It has been tested with great thoroughness and it has passed every test. Its landmark successes include predicting the existence of black holes and the recent discovery of gravitational waves. The other great theory is the standard model of particle physics, which is a fantastically successful theory of the structure of matter and fundamental particles. The standard model has also passed every test with great precision; its most recent major success being the discovery of the Higgs boson by the Large Hadron Collider (LHC) in 2012 completing the standard model table of particles. The tremendous power of these great theories makes it all the more alarming that physicists and astronomers have no idea what most of the universe is made of.

Many possible explanations of dark matter have been considered, but most have now been ruled out. All the obvious answers—dark gas clouds that have not yet formed stars or burnt-out stellar remnants such as white dwarfs and neutron stars—cannot be correct. Likewise, supermassive black holes are big, but they only make up a small fraction of a percent of the mass of a galaxy. There cannot be sufficient mass in the form of black holes to account for dark matter. Our models of how the lightest atoms, deuterium, helium, and lithium, were created in the immediate aftermath of the Big Bang work very well,[18] and their agreement with observations would be ruined if the universe had formed with much more ordinary matter. This suggests that dark matter must consist of some unknown type of matter, totally distinct from all the matter with which we are familiar. The consensus among physicists is that dark matter can only be explained by the presence of vast quantities of unknown particles that have existed since the early universe. These hypothetical particles are often referred to as WIMPs (Weakly Interacting Massive Particles). Intense efforts are being made to track them down.

There are two main approaches. One is to attempt to create WIMPs in the laboratory. This is the domain of the LHC which is designed to collide ultra-high-energy proton beams and examine the resulting particle debris. Physicists are on the lookout for any new particles created in the LHC that might explain dark matter. The other approach is to wait for the WIMPs to come to us. The Earth is continually bombarded by cosmic rays, which are high-energy particles that arrive from deep space. There are numerous sites around the world where physicists are hoping to detect signs of dark matter particles in cosmic rays. These include the LUX-ZEPLIN experiment deep underground at the Sanford Underground Research Facility in South Dakota in the United States (Figure 9.24). LUX-ZEPLIN is an international project involving scientists from around the world. At the heart of the experiment is a tank containing 7 tonnes of ultra-pure liquid xenon. If WIMPs exist, occasionally one should arrive as a cosmic ray and hit the nucleus of a xenon atom in the tank. As the nucleus recoils it releases a brief burst of light. Some of the electrons in the xenon atom are shaken loose in the collision and these electrons are attracted to the top of the tank by an electric field where they collide with a thin layer of gas atoms and emit a second pulse of light. The light pulses are detected using photo-multiplier tubes and their properties, such as the time delay between the two pulses, can be used to deduce the characteristics of the cosmic ray particle. As

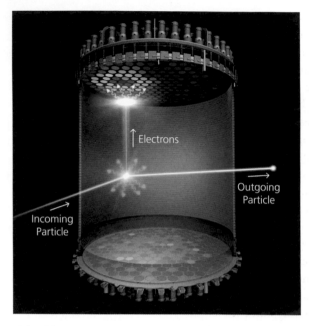

Figure 9.24 The LUX-ZEPLIN experiment. Schematic representation of how an incoming cosmic ray particle is detected by the experiment.

yet, no dark matter particles have been detected. So, despite the efforts of the LUX-ZEPLIN team and the work of scientists at other research facilities around the world, the identity of dark matter remains a mystery.

The Dark Side of the Universe

The new star whose appearance shocked Tycho Brahe in 1572 was a Type Ia supernova. As discussed in Chapter 7, such supernovae occur when white dwarf stars blast themselves apart. These eruptions are extremely bright, and they all have a similar intrinsic luminosity so they are useful standard candles for probing the furthermost depths of the universe. In the 1990s two research teams, the High-Z Supernova Search Team and the Supernova

Cosmology Project, were set up to study Type Ia supernovae in extremely distant galaxies with the aim of refining large-scale distance measurements. The results of these collaborations came as an enormous surprise to astronomers and physicists when they were announced in 1998. It had been assumed that the rate at which the universe is expanding must gradually decrease as time passes due to the gravitational attraction of all the matter within the universe. But the supernova studies indicate that the rate of expansion of the universe is increasing. This means there is an unknown force, similar to that described by Einstein's cosmological term, counteracting gravity and pushing the universe apart. It has been named *dark energy* by analogy with dark matter. But whereas dark matter received its name because it does not emit light, dark energy is *dark* because its origin is a complete mystery.

Some physicists have cast doubt on these results, as dark energy has no explanation within established physics, and the evidence relies on observations of the furthest outposts of the universe stretching current technology to its limits. However, analysis of the cosmic microwave background has added weight to the supernova studies, confirming the existence of dark energy. The accelerating expansion of the universe due to the effect of dark energy is included in Figure 9.25 which is a schematic representation of the evolution of the universe.

The Flat Universe Society

When we look out into the heavens, we are seeing light that set out towards us long ago. We can peer no further than 13.8 billion light years in any direction. This is the distance to our cosmological horizon which is defined by the cosmic microwave background. Light from any further away could not have arrived here since the origin of the universe.

If the universe is expanding like a balloon, then we would expect it to appear curved, like the surface of a balloon,[19] but the entire expanse of space that we can see looks flat. Should we all join the Flat Universe Society? Perhaps not; we are familiar with the fact that the Earth looks flat in our vicinity, even though we know it is spherical. Similarly, it seems that when we gaze into the furthest depths of space, we are seeing just a small portion of the whole cosmos. There is every reason to assume that the universe continues onwards well beyond the horizon and is therefore very much larger than the region we can see.

The microwave background is remarkably uniform from horizon to horizon. It is the same temperature throughout the sky. Yet the microwaves from one direction have travelled 13.8 billion light years, as have those from the opposite direction; they originate from two regions that are almost 28 billion light years apart. This is a little perplexing. Imagine heating a bowl of porridge in a microwave oven. There may be lumps in the porridge and it may not be heated evenly but if we leave it for a couple of minutes the heat will spread throughout the porridge until it is the same temperature throughout. Like our porridge, the temperature of the microwave background is completely uniform. But,

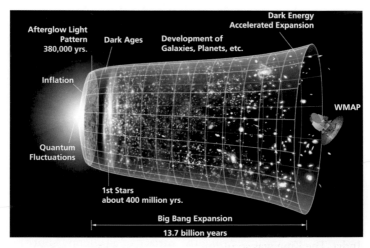

Figure 9.25 Schematic representation of the history of the universe.

unlike our porridge, there has not been sufficient time since the dawn of creation for the universe to achieve a uniform temperature. No radiation or other communication has passed between two regions on opposite horizons of the sky, yet they have the same temperature.

This might sound like a minor philosophical conundrum. It is not the sort of problem that would keep most people awake at night. Cosmologists are light sleepers, however. Just like in the story of the Princess and the Pea, any lumpy piece of universe under the mattress will keep a cosmologist awake all night.

The Grand Vizier's Garden Party

> Her words, when they had ceased, were greeted by
> A sparkling of scintillas in the spheres,
> As showers of sparks from molten metal fly.
> Tracing each fiery circle that was theirs,
> They numbered myriads more than the entire
> Progressive doubling of the chess-board squares.
> DANTE ALIGHIERI, *Paradise* XXVIII (88–93)[20]

In 1981 an American cosmologist, Alan Guth (1947–), proposed a solution to these issues. He suggested that during the very, *very* early epoch immediately after the Big Bang, for just the tiniest fraction of a second, the universe inflated exponentially. In the most minuscule instant, the universe enlarged from an infinitesimal speck to perhaps the size of a pea. It then continued to expand at a much more leisurely constant pace until reaching the size that we observe today.

This reminds me of a famous story about the origin of the game of chess, as told by the thirteenth-century Kurdish writer Ibn Khallikan.[21] According to the tale, chess was invented by the Grand Vizier Sissa ben Dahir, who offered it to his lord, the King of India. The King was so pleased with the game that he asked Sissa to name his reward. Sissa's reply was that the King could repay him by placing a grain of wheat on the first square of the chess-board, two grains on the second square, four on the next, and on each subsequent square, twice the previous one. The King readily agreed to this modest-sounding request. Only later would he realize that he had given away wheat worth far more than his entire kingdom.

The story of the Grand Vizier's reward illustrates the deceptive power of exponential growth. We can quickly work out how many grains there are on the chessboard. There is one grain on the first square. Then doubling on each subsequent square gives us: 2, 4, 8, 16, 32, 64, 128, 256, 512, 1024 grains. Rather conveniently $2^{10} = 1024$; this is the number of grains on the 11th square of the board. We can call this 1000 and represent it by the letter 'k' (i.e. $2^{10} \sim 1000 = k$), then continue piling up grains on each square, doubling in units of k.

After ten more doublings, we reach: $2^{10} \times k = k^2 = M = 2^{20}$, where 'M' represents a million; this is the number of grains on square 21. Ten more doublings give: $2^{10} \times M = G = 2^{30}$, where 'G' represents a billion; this is the number of grains on square 31. Onwards and upwards: $2^{60} = 2^{30} \times 2^{30} = G^2 = Q$, where Q represents a billion billion or, in other words, a quintillion; this is the number of grains on square 61. This doubles to 2Q on square 62, 4Q on square 63, and finally 8Q, or 8 quintillion, on the last square. The sum of all the grains on the board is twice this, giving a grand total of 16 quintillion.[22]

To simplify the calculation, we approximated 2^{10} as 1000, so we have not found the exact number of grains of wheat, but that matters little when it comes to an exponential explosion; the king is bankrupt whether we use the exact figure or not. The exact figure for the grand total is

$$2^{64} - 1 = 18,446,744,073,709,551,615,$$

almost 18.5 quintillion.

According to Alan Guth, the universe underwent just this sort of exponential growth in its first moments. The idea is that if we go back to an almost infinitesimal thousandth of a quintillionth of a quintillionth of a second, or 10^{-39} seconds, after the Big Bang, the universe was doubling in size every 10^{-39} seconds. According to the inflationary model the universe doubled in size about sixty times,[23] which parallels the chess board story quite nicely. After sixty doublings, the universe would have inflated by a factor of a quintillion but, if we check our watch, the time is now just 60×10^{-39} seconds after the Big Bang. The universe has blown up out of all proportion, but we are still within the first 10^{-37} seconds of the beginning. At this point inflation is turned off and the universe expands at a constant rate, such that after 10 billion years the radius of the universe is ten times its radius after 1 billion years, for instance.

Guth suggested that a very early inflationary epoch would explain the flatness problem by enlarging the universe well beyond the horizon that we can see. It would also solve the uniform temperature issue because, in this model, the temperature could have equalized when the universe was just a mote in God's eye, before it underwent its exponential expansion.

Figure 9.25 shows a schematic picture of the universe that was produced to commemorate the analysis of the microwave background by NASA's WMAP. The image depicts the Big Bang as a flash, followed by the brief inflationary era, then a dark era before the first stars formed. Subsequently galaxies evolve and we arrive at the universe as we know it today. In the image, the universe is represented as shaped like a bell resting on its side. This is to account for the initial phase of extremely rapid inflationary expansion, the long ages of steady expansion, as discovered by Edwin Hubble, and the accelerated expansion due to dark energy, which produces the outward curving of the bell's lip. It is worth noting that the image is only intended to be suggestive. The radius of the universe in the current epoch, which corresponds to the open end of the bell, is over 1000 times its radius when the microwave background was emitted.

Physicists are divided about the merits of cosmological inflation as there does not seem to be a natural way to turn inflation on and off at the appropriate times within the known framework of fundamental physics. This has led to accusations that inflation simply transforms features of the present-day universe that we don't understand into issues in the first instant of Creation that we don't understand; sweeping the problems of cosmology under the carpet with an extremely long broom. So is there any observational evidence for inflation that would settle the issue?

Inflation is a theory of what happened less than a quadrillionth of a quadrillionth of a second after the Big Bang. For decades it appeared to be the sort of cosmological speculation that could never be tested. But there is a signature that might allow inflation to be revealed. The surging inflationary expansion of the universe would have generated gravitational waves, leaving the entire universe ringing like a bell. And this stretching and squeezing of space would etch an indelible mark on the cosmic microwave background radiation. A team of American scientists are searching for relics of these ancient shockwaves in the fabric of space with a telescope known as BICEP (Background Imaging of Cosmic Extragalactic Polarization) that has been constructed at the South Pole. BICEP operates at just 4 degrees above absolute zero and is designed to study the microwave background.

In March 2014, the BICEP team announced that they had seen evidence for cosmological inflation. Unfortunately, these claims were premature and were soon retracted. The erroneous microwave signal picked up by BICEP is now thought to be due to dust within our own galaxy. Since 2014 the BICEP observatory has undergone a series of upgrades and the team are confident that they will soon achieve the sensitivity required to detect the imprint of primordial gravitational waves magnified by the exponential expansion of the very early universe. So the day of reckoning for inflationary cosmology may soon be at hand. The exploration of the earliest moments of the universe is no longer a distant dream. We could be on the verge of discovering the secrets of the first instant of creation and finding out how it all began.

BANG!

Further Reading

Chapter 2 includes material that I originally wrote as 'Venus in the Face of the Sun' in the June 2012 issue of *History Today*.

I have written a number of articles about fundamental physics for my blog, which can be found on the Quantum Wave website at http://www.quantumwavepublishing.com.

The following books are highly recommended sources of information about gravity and some of the other topics mentioned in this book.

Alan H Guth, *The Inflationary Universe: The Quest for a New Theory of Cosmic Origins* (London: Vintage, 1998).
Guth is the architect of the idea of an inflationary expansion of the extremely early universe. This is an accessible account of this controversial idea written for the non-specialist reader.

Clifford M Wills and Nicolás Yunes, *Is Einstein Still Right?: Black Holes, Gravitational Waves, and the Quest to Verify Einstein's Greatest Creation* (Oxford: Oxford University Press, 2020).
Clifford Wills is acknowledged to be the world's leading authority on testing Einstein's theory of general relativity. *Is Einstein Still Right?* is an enthralling and up-to-date survey of the many tests to which the theory has been subjected.

Jo Marchant, *Decoding the Heavens: Solving the Mystery of the World's First Computer* (London: Windmill, 2009).
Decoding the Heavens is an account of the discovery of the Antikythera mechanism and the painstaking detective work that has revealed how it worked and has enabled its reconstruction.

Julian Barbour, *The Discovery of Dynamics* (Oxford: Oxford University Press, 2001).
Barbour describes the historical development of dynamics, from antiquity up to the time of Newton. The book contains valuable insights into the work of Hipparchus, Ptolemy, Copernicus, Kepler, Galileo, and Newton, including many significant details that are overlooked in other accounts.

Mitchell Begelman and Martin Rees, *Gravity's Fatal Attraction: Black Holes in the Universe* 3rd edition (Cambridge: Cambridge University Press, 2020).

This is a wonderful account of the astrophysics of black holes and the observational evidence for their existence.

Nicholas Mee, *Higgs Force: Cosmic Symmetry Shattered* (London: Quantum Wave, 2012)

Gravity: From Falling Apples to Supermassive Black Holes is intended as a companion to my first book *Higgs Force*. If you enjoyed reading this book, I am sure that you will find *Higgs Force* equally rewarding.

Peter Aughton, *The Transit of Venus: The Brief, Brilliant Life of Jeremiah Horrocks, Father of British Astronomy* (Lancaster: Carnegie Publishing, 2012).

This is a great biography of a neglected figure of British science who deserves to be much more well known.

Richard S Westfall, *Never At Rest: A Biography of Newton* (Cambridge: Cambridge University Press, 1980).

Never At Rest is the definitive scientific biography of Newton. It is a work on an epic scale, as befits an account of Newton's life and work.

Walter Isaacson, *Einstein: His Life and Universe* (New York, NY: Simon and Schuster, 2007).

This is a very readable biography of Einstein that balances information about his personal life with good explanations of his scientific work.

Notes

Can You Feel the Force?

1. http://www.youtube.com/watch?v=5C5_dOEyAfk.
2. Brian Cox gives a more recent and much clearer demonstration that different masses fall with the same gravitational acceleration in a vacuum in the following video: https://www.youtube.com/watch?v=E43-CfukEgs.

Chapter 1

1. Richard Foster, *Patterns of Thought: The Hidden Meaning of the Great Pavement of Westminster Abbey* (London: Jonathan Cape, 1991) Chapter 5: The Inscription.
2. Foster, *Patterns of Thought*, Chapter 5: The Inscription.
3. The emphasis on the number three in the rhyme may also derive from Aristotle, who wrote in the opening section of *On the Heavens*:

   > *A magnitude if divisible one way is a line, if two ways a surface, and if three a body. Beyond these there is no other magnitude, because the three dimensions are all that there are, and that which is divisible in three directions is divisible in all. For, as the Pythagoreans say, the world and all that is in it is determined by the number three, since beginning and middle and end give the number of an 'all', and the number they give is the triad. And so, having taken these three from nature as (so to speak) laws of it, we make further use of the number three in the worship of the Gods.*

   Aristotle, *On the Heavens*, translated by J. L. Stocks, Princeton, NJ: Princeton University Press, 1985, https://www.degruyter.com/document/doi/10.1515/9781400835843-011/ html.
4. Following the philosopher Empedocles.
5. Bertrand Russell, *A History of Western Philosophy*, Chapter XXIII Aristotle's Physics. Routledge: London, 2004, p.213.
6. Within the framework of the Hierarchic Universe all spheres of knowledge were incorporated into a grand unified theology. In the realm of biology the hierarchy took the form of the Great Chain of Being. Every organism from the lowliest worm to the higher mammals had its God given place in the order of things. The pinnacle of the terrestrial hierarchy was, of course, man himself, but even

within the human species every grade of person, peasant, merchant, priest, lord, took their place in a linear procession. The king stood at the very tip of the spire, the instrument of God's will on Earth. The hierarchy continued beyond human beings, through the various ranks of angels in their allotted heavenly spheres, onwards to its ultimate omniscient terminus, God. And downwards from the lowliest human to the higher animals and on to the most primitive creatures. Thus Mankind, with its divine intellect trapped in its sensuously brutish body, held the intermediate station halfway between the worm and God. Even the vermian end of the hierarchy was extended to inanimate matter through symbolic correspondences and sympathies, such as those between precious stones and the planets.

7. The late medieval church had an overwhelming grip on the whole of European society. The Christian world view of that time was so alien to our own in this age of science that it is very hard to imagine how its educated citizens must have thought. Their world was interpreted in terms of symbolism, the main function of the objects within it was considered to be to symbolically represent the entire range of religious mysteries. In this climate scientific discoveries could not occur. If a phenomenon is not important in itself, but only in so far as it is a metaphor for a deeper mystery, it is not going to be critically analysed to uncover its own inner workings. The belief in mystical and symbolic correspondences between objects also precludes the observation of genuine causal relationships. The cosmogony of the medieval church was clearly never envisaged as a model of the real universe, but that was not the criterion by which it would have been judged. The physical world was not seen as being important.

8. http://everything2.com/title/Dante%2527s+use+of+threes+in+the+Inferno.

9. Dante Alighieri, *Paradise*, Canto XXXIII 133, translated by Dorothy L. Sayers (London: Penguin, 1949).

10. In Book XI of *The Republic*, Plato offers a poetic description of the cosmos. He suggests that the planets form concentric circles with the earth motionless at the centre. These planetary orbs are depicted as eight whorls wound around the diamond shaft of the Spindle of Necessity. Moving outwards from the centre the whorls represent the orbits of the moon, the Sun, each of the planets, and the fixed stars. The motion of the planets on these eight circles produces eight notes of constant pitch composing a single scale. Outside

the last sphere Plato imagined the three Fates, Lachesis, Clotho, and Atropos, daughters of Necessity, bethroned equidistant from the centre, singing of the past, present, and future respectively, while giving the occasional twirl to the spindle.

11. This is because it takes Venus 224.7 days to orbit the sun, which means that in 8 years Venus orbits the sun almost exactly 13 times. So the positions of the Earth and Venus in their orbit will have returned to almost exactly the same positions.

12. NASA Catalogue of Solar Eclipses, http://eclipse.gsfc.nasa.gov/SEsaros/SEsaros139.html.

13. Every third eclipse in a saros cycle will be visible from the same general region of the globe. Taking every third eclipse gives us the Callippic cycle which is three times as long as the saros cycle, i.e. fifty-four years and thirty-four days.

14. Herodotus, *The Histories*, vol 1, 1.74.2 (George Rawlinson tr, London: Dent, 1910) 136; Mark Littmann, Ken Willcox, and Fred Espenak, *Totality: Eclipses of the Sun* (Oxford: Oxford University Press, 2009) p.48, gives the date of the eclipse predicted by Thales.

15. 'When, in the sixth year they encountered one another, it so fell out that, after they had joined battle, the day suddenly turned into night. This transformation of day into night was foretold to the Ionians by Thales of Miletus.' Herodotus, *The Histories* I (London, Penguin, 2003) 74.

16. This story in Herodotus may have been the inspiration behind a very similar incident that occurs in Mark Twain's novel, *A Connecticut Yankee in King Arthur's Court* (London, Collins Classics, 2012).

17. Half a saros interval prior to a solar eclipse, i.e. 9 years 5.5 days, a lunar eclipse occurs with similar properties to the solar eclipse. Thales may have known of the solar eclipse of 18 May 603 BC, which would have been visible in the Middle East, and the lunar eclipse of 24 May 594 BC.

18. Jo Marchant, *Decoding the Heavens: Solving the Mystery of the World's First Computer* (London: Windmill, 2009).

19. The night sky is like a huge sphere, and astronomers indicate the positions of stars by their coordinates on this sphere. This is very like the mapping of places on Earth by their longitude and latitude. Any place can be located by providing two angles that determine the position with respect to two circles around the Earth—one being the equator and the other being, quite arbitrarily, the Greenwich

meridian (a meridian is a circle passing through the North and South Poles, so the Greenwich meridian is the circle that passes through the North and South Poles and Greenwich). Greenwich is defined to be zero longitude. The point at which the Greenwich meridian crosses the equator is defined to be the origin of the coordinate system—in other words, the point which has coordinates (0° longitude, 0° latitude). To specify the position of any other place on Earth, we give an angle representing the longitude of the place and an angle representing the latitude of the place. To find a place on the globe from its longitude and latitude, we start at the origin of the coordinate system and move around the equator an angle corresponding to the longitude, which can be any angle between zero and 360°, then we move around a meridian northwards or southwards by an angle corresponding to the latitude. Latitude is specified by an angle between –90° and +90° or, alternatively, by an angle between 0° and 90° south or an angle between 0° and 90° north. Astronomers specify the position of a star in very much the same way, by projecting this coordinate system onto the sphere of the heavens, so the position of any star or other object is determined by two angles. The apparent distance in the night sky between any two celestial objects is given as an angle, which would be the angle by which the celestial globe would need to be turned along an arc joining the positions of the two objects to bring them together.

20. We can think of a star as sitting in the centre of an enormous circle with the Earth positioned on the circumference of the circle. To determine the distance to the star we must work out the radius of this circle, which is the circumference divided by 2π. As the Earth orbits the Sun, its position changes. In January, it will be at one point on the circumference. Six months later, it will have reached a slightly different position on the circumference. The two viewpoints of the star are the ends of one edge of a polygon inscribed within the circle. The length of this edge is the diameter of the Earth's orbit (or twice the distance to the Sun). Imagine that the star's position shifts by 1° or 1/360th of a circle. Then this polygon is a 360-gon. The polygon very closely approximates a circle, as it has so many sides, and its perimeter is almost the same as the circumference of the circle. So, for a star whose position shifts by 1° during the course of a year, 360 times the diameter of the Earth's orbit is a very good approximation

to the circumference of the circle with the star at the centre. If we approximate 2π as 6, which is close enough for our purposes, then the distance to the star is about $360/6 = 60$ times the diameter of the Earth's orbit.

21. The size of the retrograde loops for each planet is partially due to parallax but, as the planets are moving in the same direction as the Earth, the loops are not as big as they would be if the planets were stationary relative to the Earth. The sizes of the retrograde loops are approximately as follows: Jupiter: $10°$, Saturn: $6°$, Uranus: $4°$, Neptune: $3°$.

22. 2×313.6 mas (milli-arcseconds) to be precise: http://en.wikipedia. org/wiki/61_Cygni.

23. The modern figure is 11.4 light years.

24. It is possible to estimate the distance to closer astronomical objects, such as the Moon or a comet, by using the Earth's diameter as the baseline and taking measurements as the Earth rotates during the course of a single night. The size of the Earth's diameter has been known since ancient times.

25. Quotation appears in Michael J Crowe, *Theories of the World from Antiquity to the Copernican Revolution* (London: Dover, 2001).

Chapter 2

1. Max Caspar, *Kepler* (London: Dover, 1993) p.59.

2. More precisely Jupiter's orbital period is 11.86 years and Saturn's orbital period is 29.46 years.

3. J.V. Field, *Kepler's Geometrical Cosmology* (London: Athlone, 1988).

4. Malcolm Longair, *Theoretical Concepts in Physics* (Cambridge: Cambridge University Press, 2003) p.26. According to Malcolm Longair, Kepler's polyhedral model fits the planetary orbits to within about 5% accuracy.

5. Arthur Koestler, *The Sleepwalkers* (London: Penguin, 1959) p.271.

6. A detailed account of this incident was recorded by Pierre Gassendi in 1654. According to Gassendi:

> *On the 10th December 1566, there was a dance at Lucas Bacmeister's house in the connection to a wedding. Lucas Bacmeister was a professor of theology at the university of Rostock, where Tycho studied. Among the guests were Tycho Brahe*

and another Danish nobleman, Manderup Parsberg. They started an argument and they separated in anger. The 27th of December, this argument started again, and in the evening of the 29th of December a duel was held. It was around 7 in the evening and in darkness. Parsberg gives Tycho a cut over his nose that took away almost the front part of his nose. Tycho had an artificial nose made, not from wax, but from an alloy of gold and silver[*], and put it on so skillfully, that it looked like a real nose Wilhelm Janszoon Blaeu, who spent time with Tycho for nearly two years, also said that Tycho used to carry a small box with a paste or glue, with which he often would put on the nose.

https://lost-contact.mit.edu/afs/nada.kth.se/home/ass/fred/public_html/tycho/nose.html.

7. See Nicholas Mee, *Higgs Force: Cosmic Symmetry Shattered* (London: Quantum Wave, 2012).

8. Noel M. Swerdlow, 'Astronomy in the Renaissance' in Christopher Walker (ed.), *Astronomy Before the Telescope* (London: British Museum Press, 1999) p.210.

9. Longair, *Theoretical Concepts in Physics*, p.23.

10. https://lost-contact.mit.edu/afs/nada.kth.se/home/ass/fred/public_html/tycho/nose.html.

11. Longair, *Theoretical Concepts in Physics*, p.23.

12. https://www.bbc.co.uk/news/11756077#:~:text=The%20body%20of%20a%2016th,the%20Bohemian%20Emperor%20Rudolf%20II.

13. Joshua Gilder and Anne-Lee Gilder, *Heavenly Intrigue: Johannes Kepler, Tycho Brahe, and the Murder Behind One of History's Greatest Scientific Discoveries* (New York, NY: Doubleday, 2004).

14. Spiegel International, http://www.spiegel.de/international/europe/0,1518,601729-2,00.html.

15. A good place to look is the website *Heavens Above* (http://www.heavens-above.com). This website provides information about all the planets, the positions of bright comets, and much else besides.

16. Nicholas Mee, *Higgs Force: Cosmic Symmetry Shattered* (London: Quantum Wave, 2012).

17. Which is longer—summer or winter? If you live in Britain, you may well feel that winter is definitely longer than summer. However, in the northern hemisphere, summer is actually over five and a half days longer than winter. The year can be divided into four

quarters by the equinoxes and the solstices. The equinoxes are the days on which the Sun is directly overhead at the equator, which means that at this time of year, the day is divided in half, with the period of daytime equalling the period of night-time. This is the origin of the name 'equinox' (equal night). The solstices are the days on which the Sun is overhead at the tropics of Cancer or Capricorn, its furthest excursion northwards or southwards respectively. These are the days on which the apparent journey of the Sun northwards or southwards stops before the Sun begins its return to the other hemisphere. Hence, the name 'solstice' which means stationary Sun. Officially, the seasons are designated as the periods between these four marker days. The Earth is closest to the Sun on 3 January and the Earth moves faster around its orbit when it is closer to the Sun. This means that the period of time between the winter solstice (21 December) and the spring equinox (21 March) is shorter than the period of time between the summer solstice (21 June) and the autumn equinox (21 September). Wikipedia entry on season duration, http://en.wikipedia.org/wiki/File:SeasonDuration.png.

18. One complete rotation of the Earth takes around 23 hours and 56 minutes. This is the time that it takes for a star that is due south in the sky to return to the same position due south. On average, it takes 24 hours for the Sun to return to the same position in sky. The extra four minutes are due to the motion of the Earth around its orbit. In a year, these four minutes add up to one complete rotation. Thus, in a year—the time taken to complete one orbit—the Earth rotates on its axis $366\frac{1}{4}$ times.

19. More precisely, he determined it to be is $1°50'$ (one degree and 50 minutes).

20. Julian Barbour, *The Discovery of Dynamics* (Oxford: Oxford University Press, 2001).

21. Selections from Kepler's *Astronomia Nova: A Science Classics Module for Humanities Studies*, selected, translated, and annotated by William H. Donahue (Mineola, NY: Green Lion Press, 2008) p.256.

22. Quoted in Max Caspar, *Kepler* (London: Dover, 1993) p.128.

23. Quoted in Caspar, *Kepler*, p.128.

24. The orbit that Kepler had determined for the Earth was a circle, with the Sun off-set from the centre of the circle. This is a very good approximation for the Earth's orbit, as the eccentricity of the Earth's

orbit is small. Kepler would later show that, like Mars, the Earth's orbit is actually elliptical.

25. In 1604, Kepler published *Astronomiae Pars Optica (The Optical Part of Astronomy)* Thomas Little Heath tr, Carruthers Press, 2015.

26. The trajectory of Mars through space could now be described quite simply. All that was required to specify it precisely were the size of the orbit, the eccentricity of the orbit, and the direction of the axis of the ellipse, corresponding to the direction from the Sun towards the position of the closest Martian approach to the Sun and represented by a point on the celestial sphere.

27. In Ptolemy's model, the Earth lies on one side of the centre of a planet's orbit and the equant point is an equal distance on the opposite side. The same construction in the Keplerian system would put the Sun at one focus of the elliptical orbit and the equant point at the other focus. To a good approximation, first order in the eccentricity, the angular velocity of a planet is constant when viewed from the second focus of the ellipse. This means that the equant point really does provide a fairly accurate way to model the motion of the planets, and this is why the Ptolemaic system worked as well as it did.

28. The orbits of the planets are ellipses, as Kepler discovered. Strictly speaking it is the semi-major axis of the planet's elliptical orbit that should be used in the harmonic law, not the radius of the orbit.

29. *Mensus eram coelos, nunc Terrae metior umbras.*
    Mens coelestis erat, corporis umbra jacet.
    KGW 19 393. MacTutor, Quotations, Johannes Kepler, http://www-history.mcs.st-and.ac.uk/Quotations/Kepler.html.

Chapter 3

1. Galileo's *The Starry Messenger*— http://www.bard.edu/admission/forms/pdfs/galileo.pdf.

2. It may be significant that Galileo was observing a swinging chandelier in a cathedral, because such a pendulum would no doubt have been very long and the amplitude of its oscillations rather small. The period of oscillation of such a pendulum would then have been constant to a good approximation, as Galileo observed.

3. Julian Barbour, *The Discovery of Dynamics* (Oxford: Oxford University Press, 2001) p.356.

4. Galileo did not quite have the same concept of inertia as Newton. He believed that the motion of the ball would not be a straight line, but would follow the surface of the Earth, i.e. it would be circular. (Barbour, *The Discovery of Dynamics*).

5. The passage continues:

> In jumping, you will pass on the floor the same spaces as before, nor will you make larger jumps toward the stern than toward the prow even though the ship is moving quite rapidly, despite the fact that during the time that you are in the air the floor under you will be going in a direction opposite to your jump. In throwing something to your companion, you will need no more force to get it to him whether he is in the direction of the bow or the stern, with yourself situated opposite. The droplets will fall as before into the vessel beneath without dropping toward the stern, although while the drops are in the air the ship runs many spans. The fish in their water will swim toward the front of their bowl with no more effort than toward the back, and will go with equal ease to bait placed anywhere around the edges of the bowl. Finally, the butterflies and flies will continue their flights indifferently toward every side, nor will it ever happen that they are concentrated toward the stern, as if tired out from keeping up with the course of the ship, from which they will have been separated during long intervals by keeping themselves in the air. And if smoke is made by burning some incense, it will be seen going up in the form of a little cloud, remaining still and moving no more toward one side than the other. The cause of all these correspondences of effects is the fact that the ship's motion is common to all the things contained in it, and to the air also. That is why I said you should be below decks; for if this took place above in the open air, which would not follow the course of the ship, more or less noticeable differences would be seen in some of the effects noted.

Galileo Galilei, *Dialogue Concerning the Two Chief World Systems*, translated by Stillman Drake (Berkeley: University of California Press, 1953) pp.186–7.

6. Barbour, *The Discovery of Dynamics*, p.371.

7. This law can be represented as: $s = \frac{1}{2} at^2$, where s is distance, a is the acceleration due to gravity, and t is time.

8. When showing that the path of a projectile is a parabola, the force of gravity is assumed to be constant which is a good approximation if the projectile's path remains close to the surface of the Earth. If the force of gravity is assumed to be an inverse square law, then the calculation would show that the path of the projectile is along an arc of a very eccentric ellipse, with a focus at the centre of the Earth.

9. *The Essential Galileo*, edited and translated by Maurice A Finocchiaro (Indianapolis, Indiana: Hackett, 2008) Chapter 10, p.295–356.

10. And lest some should persuade ye, Lords and Commons, that these arguments of learned men's discouragement at this your Order are mere flourishes, and not real, I could recount what I have seen and heard in other countries, where this kind of inquisition tyrannises; when I have sat among their learned men, for that honour I had, and been counted happy to be born in such a place of philosophic freedom, as they supposed England was, while themselves did nothing but bemoan the servile condition into which learning amongst them was brought; that this was it which had damped the glory of Italian wits; that nothing had been there written now these many years but flattery and fustian. There it was that I found and visited the famous Galileo, grown old a prisoner to the Inquisition, for thinking in astronomy otherwise than the Franciscan and Dominican licensers thought.

  John Milton, *Areopagitica*. The booklet is available in *Areopagitica and Other Writings* by John Milton (Author), William Poole (Editor), Penguin Classics (London, 2014).

11. All the planetary orbits in the solar system lie in the same plane (or nearly so). This plane also contains the Sun's equator and, when projected onto the sky, it is called the ecliptic. The paths of the planets across the sky always lie near to this circle. Although the planetary orbits all lie close to the ecliptic, their alignment is not perfect. The orbits are tilted slightly with respect to each other. The orbit of Venus is inside the Earth's orbit, and it is inclined at an angle of around three and a half degrees relative to the Earth's orbit around the Sun. This means that, as viewed from Earth, on almost all occasions when Venus overtakes the Earth on its inner track around the Sun, it passes well above or well below the Sun's disc (the apparent diameter of the Sun's disc is about half a degree).

12. The Julian calendar was still in use in Britain at this time. All the dates that are quoted from the seventeenth century were those in

use at the time. The date of the transit according to the Gregorian calendar was 4 December 1639.

13. NASA Eclipse website, http://eclipse.gsfc.nasa.gov/transit/catalog/VenusCatalog.html.

14. To be precise, 8.85 years.

15. Wikipedia entry on lunar precession, http://en.wikipedia.org/wiki/Lunar_precession.

16. Peter Aughton, *The Transit of Venus: The Brief, Brilliant Life of Jeremiah Horrocks Father of British Astronomy* (Lancaster: Carnegie Publishing, 2012; Phoenix, 2004) p.122.

17. Aughton, *The Transit of Venus*, p.128.

18. Aughton, *The Transit of Venus*, p.119.

19. This passage from one of Crabtree's letters is quoted on page 174 of *The Transit of Venus* by Aughton.

20. This quotation is not original to Newton, but it is found in a private letter from Newton to Robert Hooke and is considered by some to be a jibe at Hooke's expense, as Hooke had a short and hunched figure.

Chapter 4

1. Richard S Westfall, *Never at Rest: A Biography of Newton* (Cambridge: Cambridge University Press, 1980) pp. 402–3.

2. Newton gave this account to De Moivre long after the event. At the time of the meeting, Halley had not yet received his doctorate and Newton had not been awarded his knighthood.

3. Quoted in Westfall, *Never at Rest: A Biography of Newton*, p.143.

4. The other two are arguably the discovery of mathematical proof by the ancient Greeks and the invention of symbolic algebra.

5. Galileo believed that inertial motion would follow the curvature of the Earth, so in his view the path of a freely moving, unimpeded body would only approximate to a straight line.

6. Another example is a rocket. When a rocket is propelled upwards by the explosive expansion of the gases in its combustion chamber, the upwards force that acts upon the rocket is exactly equal to the downwards force on the exhaust gases that it emits.

7. Newton's Third Law is closely related to the Law of Conservation of Momentum.

8. It was clear to Newton that gravity acts throughout the solar system. The Sun has its system of planets, and the Earth is orbited by the Moon. Jupiter is orbited by the four moons discovered by Galileo,

and more recently five satellites of Saturn had been discovered: Titan in 1655 by Christiaan Huygens, and Rhea, Dione, Tethys, and Iapetus in 1671–1672 by Giovanni Cassini. The force of gravity appeared to diminish with distance from the giant planets in exactly same way as it diminishes with distance from the Sun. This was clear because each satellite system obeys its own version of Kepler's Third Law. The square of the orbital period of the satellites of Jupiter grows as the cube of the radius of their orbit, and similarly for the satellites of Saturn, but with a different constant of proportionality (due to the differing masses of Jupiter and Saturn). This was a great clue. It showed that the force of gravity operates in the same way everywhere it was possible to look.

9. The Sun's gravitational pull on the Earth is proportional to the mass of the Sun. The universal law of gravity implies that the Earth pulls on the Sun, just as the Sun pulls on the Earth. This is in accordance with Newton's Third Law of Motion, which requires that forces arise in equal and opposite pairs (so rather than the Earth orbiting the Sun, it is more accurate to say that the Sun and the Earth revolve around their common centre of mass). The mass of the Sun is vastly greater than the mass of the Earth, so their centre of mass is close to the Sun.

10. Edmund Halley is believed to have toyed with the idea that the Earth might be a hollow shell. Wikipedia entry for Hollow Earth: https://en.wikipedia.org/wiki/Hollow_Earth.

11. The volume of the sphere beneath us is proportional to the cube of the distance to the centre. If the density is uniform, then the mass of this sphere is also proportional to the cube of the distance to the centre. The gravitational force is inversely proportional to the square of the distance to the centre. Therefore, assuming uniform density, the force is proportional to the distance from the centre.

12. If the Earth's density were uniform, we would undergo simple harmonic motion.

13. As the Moon orbits the Earth once a month, it crosses the meridian due south in the sky every 24 hours and 50 minutes, which means that the time between high tides is around 12 hours and 25 minutes.

14. In addition to being stretched in the direction towards the Moon, the Earth is squeezed in the plane perpendicular to this direction. This is best understood in terms of Einstein's theory which is presented in

a later chapter. It means that low tides are lower than they would otherwise be.

15. American Mathematical Society, 'Fourier Analysis of Ocean Tides III'. http://www.ams.org/samplings/feature-column/fcarc-tidesiii2.

16. Anecdotes, Observations and Characters, of Books and Men Vol 1 (page 158), by the historian Joseph Spence (1820), available in an OUP edition from 1966. These words are reported to have been uttered by Newton to Chevalier Andrew Michael Ramsey just before his death.

17. Westminster Abbey website, 'Sir Isaac Newton', http://www.westminster-abbey.org/our-history/people/sir-isaac-newton.

18. Meike Molthof, 'The Industrial Revolution and a Newtonian Culture', *E-International Relations* website, 24 August 2011, http://www.e-ir.info/2011/08/24/the-industrial-revolution-and-a-newtonian-culture/.

19. Clifford Bekar and Richard Lipsey, *Science, Institutions and the Industrial Revolution* (October 2002).

20. *Candide*, Chapter VI, available in Penguin Classics (London, 2006).

21. Among Michell's achievements was his demonstration in 1750 that the magnetic force exerted by each pole of a magnet is proportional to the inverse square of the distance from the pole.

22. William Herschel was influenced by Michell's work on double stars. After Michell's death, Herschel purchased his ten-foot telescope. Although it was no longer in working condition, Herschel used it as a model for a telescope that he constructed himself. https://www.messier.seds.org/xtra/Bios/michell.html

23. 'The Country Parson Who Conceived of Black Holes', American Museum of Natural History website, http://www.amnh.org/education/resources/rfl/web/essaybooks/cosmic/cs_michell.html.

24. In 1774, the Astronomer Royal, Nevil Maskelyne, led an expedition to determine the strength of gravity by measuring the deflection of a pendulum due to the gravitational attraction of an isolated Scottish mountain known as Schiehallion. The experiment was carried out with great care and included a systematic survey of the mountain to ascertain its volume. The result produced a value for Newton's gravitational constant (or equivalently, the mass of the Earth) with an accuracy of around 20%.

25. Philip Ball, *Elegant Solutions: Ten Beautiful Experiments in Chemistry* (London: Royal Society of Chemistry, 2005) p.26.

26. This could be deduced from Kepler's Third Law of Planetary Motion, which relates the size of the orbit to its period.

27. Tom Standage, *The Neptune File* (London: Penguin, 2001).

28. Standage, *The Neptune File*, p.60.

29. Standage, *The Neptune File*, p.108.

30. This quote is from 'Charlie X', the second episode of the first season of the original Star Trek. Written by Dorothy C. Fontana from a story by Gene Roddenberry, and directed by Lawrence Dobkin, it first aired on September 15, 1966.

31. Clifford M Will, *Was Einstein Right?* (Oxford: Oxford University Press, 1986) p.91. (These are modern figures and are more accurate than those used by Le Verrier.)

32. The quote is from Wordsworth's poem The Prelude. It is from Book 3: Residence in Cambridge – lines 30-35 and can be found here: https://www.bartleby.com/270/1/71.html It is also found in *The Prelude: The Four Texts (1798, 1799, 1805, 1850)* (Penguin Classics), (London, 1995).

Chapter 5

1. Nicholas J Mee, *Higgs Force: Cosmic Symmetry Shattered* (London: Quantum Wave, 2012).

2. For example, Russell's teapot.

3. Galileo's understanding of inertia was not quite correct. He did not realize that an object will continue in a straight line unless acted on by a force, as stated in Newton's First Law. He believed the motion of an unperturbed body would be circular—it would revolve around the Earth. Descartes was probably the first to correctly define inertial motion.

4. Albert Einstein, 'On the Electrodynamics of Moving Bodies', 30 June 1905. *Annalen der Physik* 17 (1905): 891–921.

5. Einstein, 'On the Electrodynamics of Moving Bodies'.

6. Einstein, 'On the Electrodynamics of Moving Bodies'.

7. The experiment by Rossi and Hall was performed at an elevation of 3230 metres on Mount Evans, which is near Echo Lake, Colorado.

8. The full relativistic formula is: Inertial mass = $m(1-v^2/c^2)^{\frac{1}{2}}$. At velocities that are low compared to the speed of light, v/c is much less than 1. The formula can be expanded in powers of v/c to give: Inertial mass = $m + \frac{1}{2} m(v/c)^2 + \ldots$, where only the first two terms have been retained. The next term is multiplied by $(v/c)^4$, which is tiny except at velocities approaching the speed of light.

9. Herbert G Wells, *The Time Machine* (London: Heinemann, 1895) p.1.

10. Hermann Minkowski, 'Space And Time', a translation of an address delivered at the 80th Assembly of German Natural Scientists and Physicians, at Cologne, 21 September 1908. In Hendrik A Lorentz, Hermann Weyl, Hermann Minkowski, et al., *The Principle of Relativity: A Collection of Original Memoirs on the Special and General Theory of Relativity* (London: Dover, 1952) p.74.

11. An example of this approach is the liquid drop model of the nucleus developed by Hans Bethe, Karl Friedrich von Weizsäcker, Robert Fox Bacher, Eugene Wigner, and Niels Bohr.

12. Einstein letter to Ehrenfest in 1916. Gino Segrè, *Faust in Copenhagen: A Struggle for the Soul of Physics and the Birth of the Nuclear Age* (London: Pimlico, 2008) p.175.

Chapter 6

1. Leon Battista Alberti's treatise, '*Della Pittura*'. See also Leon Battista Alberti, *On Painting: A New Translation and Critical Edition*, Cambridge University Press; Illustrated edition (25 July 2013).

2. Quoted in Phillip Davis and Reuben Hersh, *The Mathematical Experience* (Boston, MA: 1981). The quote is from a letter written in 1820 from Farkas Bolyai to his son, János. It is translated in Philip J. Davis and Reuben Hersh, *The Mathematical Experience* (1981), p.220, which was published in the UK by Penguin (London, 1990).

3. Bolyai never accepted that Gauss had independently pre-empted his great discovery, but there is no doubt this was true. Gauss had explored non-Euclidean geometry in his notebooks and had communicated some of his results to a select few mathematicians.

4. It is negative because $+1$ multiplied by -1 is equal to -1. Positive curvature arises when the centres of curvature are both in the same direction.

5. General Investigations of Curved Surfaces by C F Gauss (Author), Peter Pesic (Editor) (Dover Books on Mathematics) (Mineola, New York, 2005).

6. James B Hartle, *Gravity: An Introduction to Einstein's General Relativity* (Boston, Massachusetts: Addison Wesley, 2003) p.15.

7. This is the stress-energy tensor.

8. This is the Einstein tensor.

9. Abraham Pais, *Subtle is the Lord: The Science and the Life of Albert Einstein* (Oxford: Oxford University Press, 1982) p.253.

10. Letter from Karl Schwarzschild to Albert Einstein, dated 22 December 1915, in *The Collected Papers of Albert Einstein*, vol.8a, doc. No.169.

11. Alfred Eisenstaedt, 'The Early Interpretation of the Schwarzschild Solution,' in Don Howard and John Stachel (eds), *Einstein and the History of General Relativity: Einstein Studies*, vol. 1 (Boston, MA: Birkhauser, 1989) 213–34.

12. Featured in the 1973 rock musical *The Rocky Horror Show*, and its 1975 film adaptation *The Rocky Horror Picture Show*.

13. Ian R Kenyon, *General Relativity* (Oxford: Oxford University Press, 1990) p.17.

14. Clifford M Will, *Was Einstein Right?: Putting General Relativity to the Test* (Oxford: Oxford University Press, 1986) p.57.

15. C W Francis Everitt et al. 'The Gravity Probe B Test of General Relativity' (2015) 32 *Classical Quantum Gravity* 224001. https://iopscience.iop.org/article/10.1088/0264-9381/32/22/224001

16. This is because in hyperbolic space around a massive body the circumference of the orbit is less than $2\pi R$, where R is the radius of the satellite's orbit.

17. According to Newtonian gravity, satellite orbits around a spherical mass should obey Kepler's Zeroth Law (i.e. they should remain in the same plane indefinitely), and the rotation of the spherical mass has no gravitational effect. According to general relativity, satellite orbits around a rotating spherical mass will not obey Kepler's Zeroth Law. Due to the Lense–Thirring effect the plane of the orbit will gradually change.

18. Hartle, *Gravity: An Introduction to Einstein's General Relativity*, p.124.

19. Richard R Pogge, 11 March 2017, 'Real-World Relativity: The GPS Navigation System', http://www.astronomy.ohio-state.edu/~pogge/Ast162/Unit5/gps.html.

Chapter 7

1. Michael White and John Gribbin, *Stephen Hawking: A Life in Science* (London: Penguin, 1991), p.131.

2. Remarkably, white dwarfs are supported by nothing more than the Pauli Exclusion Principle which states that no two electrons can occupy the same quantum state. The electrons within a white dwarf must all have their own separate quantum state and this produces a huge resistance to their being squeezed any closer together. This is called *electron degeneracy pressure*.

3. The easiest white dwarf to see with a telescope is a member of a triple star system known to astronomers as *omicron 2 Eridani*. This star system is a close stellar neighbour at about 16 light years distance. It is also known by the name Keid, which is derived from the Arabic for broken eggshell. The main star, which is visible to the naked eye, is orbited by a binary that requires a telescope to be seen. The binary consists of a white dwarf and an even fainter red dwarf. (The red dwarf is a ordinary low-mass star that is generating energy by converting hydrogen into helium.) According to Gene Roddenberry, the creator of *Star Trek*, Spock's home planet Vulcan orbits the star Keid A. Wikipedia entry for Vulcan (*Star Trek*) http://en.wikipedia.org/wiki/Vulcan_(Star_Trek).

4. Jocelyn Bell Burnell, 'Little Green Men, White Dwarfs or Pulsars?' Presented as an after-dinner speech with the title 'Petit Four' at the Eighth Texas Symposium on Relativistic Astrophysics and published in (1977) Dec(302) *Annals of the New York Academy of Science* 685–9.

5. Mitchell Begelman and Martin Rees, *Gravity's Fatal Attraction: Black Holes in the Universe*, 2nd edn (Cambridge: Cambridge University Press, 2010).

6. However, there is not necessarily anything unusual about the structure of space in the region of the event horizon.

7. 'Ivor Robinson, Founding Leader of Math, Physics Department, Dies', News Center, University of Texas at Dallas website, https://news.utdallas.edu/campus-community/ivor-robinson-founding-leader-of-math-physics-depa/.

8. https://heasarc.gsfc.nasa.gov/docs/uhuru/uhuru.html

9. James C Miller-Jones, Arash Bahramian, Jerome A Orosz et al. 'Cygnus X-1 Contains a 21-Solar Mass Black Hole—Implications for Massive Star Winds' (2021) 371(6533) *Science* 1046–49. https://science.sciencemag.org/content/371/6533/1046.

10. Quoted in Begelman and Rees, *Gravity's Fatal Attraction: Black Holes in the Universe*, 2nd edn, p. 224.

11. CP Snow's address continues:
   I now believe that if I had asked an even simpler question—such as, What do you mean by mass, or acceleration, which is the scientific equivalent of saying, Can you read?—not more than one in ten of the highly educated would have felt that I was speaking the same language. So the great edifice of modern physics goes up, and the

majority of the cleverest people in the western world have about as much insight into it as their neolithic ancestors would have had.

12. The First Law was actually formulated after the Second Law, but it is logically the first.

13. To clarify what is happening, Carnot defined a quantity S that is equal to an amount of heat Q at a specified temperature T, so

$$S = Q/T.$$

Carnot referred to this quantity as *entropy*. (Strictly speaking, this is just a contribution to the total entropy of the object so it is usually designated ΔS, and the total entropy is designated S.)

For a given amount of heat, the entropy is smaller at a high temperature than at a low temperature. If T_{high} represents a high temperature and T_{low} represents a low temperature, then

$$Q/T_{high} < Q/T_{low}.$$

(This is true because T_{high} is bigger than T_{low}, so when we divide by T_{high} we are dividing by a bigger number.) If S_{high} represents the entropy at the higher temperature and S_{low} represents the entropy at the lower temperature, then

$$S_{high} < S_{low}.$$

This means that if an amount of heat Q is transferred from a hot object to a cold object, the total entropy will increase. But if heat were transferred from a cold object to a hot object, then the entropy would decrease. We are very familiar with the first of these processes, but the second process never happens. Carnot codified this feature of the universe in the statement that, in any allowed process, the total entropy must increase.

14. Photons cannot be confined within a black hole if their wavelength is comparable to the radius of the event horizon. If M is the mass of the black hole, c is the speed of light, G is Newton's constant, and the photon wavelength is λ, then:

$$\lambda \sim 2GM/c^2.$$

If a black body emits photons with a typical wavelength λ, then its temperature is:

$$T \sim E_{photon}/k = (hc/\lambda)(1/k).$$

where k is Boltzmann's constant. The Hawking temperature of the black hole is therefore:

$$T_H \sim hc^3/2kGM.$$

Hawking's more precise quantum field theoretic derivation gives:

$$T_H = \hbar c^3/8\pi kGM.$$

15. This is much closer in scale to a 610-metre (2000-foot) British mountain rather than a Himalayan peak.
16. Begelman and Rees, *Gravity's Fatal Attraction: Black Holes in the Universe*, 2nd edn, p.267.
17. The Kerr solution describes rotating black holes. In this case the singularity actually forms a ring of infinite density rather than a point.
18. John Archibald Wheeler with Kenneth Ford, *Geons, Black Holes & Quantum Foam: A Life in Physics*, New York City: WW Norton, p.247.

Chapter 8

1. Spacetime is an incredibly stiff medium. It would take a pressure of 10^{43} N per metre squared to warp spacetime into a shape with a radius of curvature of one metre (Ian R Kenyon, *General Relativity* (Oxford: Oxford University Press, 1990) p.124).
2. LIGO Scientific Collaboration, 'Observation of Gravitational Waves from a Binary Black Hole Merger' 11 February 2016, https://www.ligo.org/science/Publication-GW150914/index.php.
3. Vassiliki Kalogera and Albert Lazzarini, 'LIGO and the Opening of a Unique Observational Window on the Universe' (21 March 2017) 114(12) Proceedings of the National Academy of Sciences of the United States of America 3017–25, Figure 1 https://www.pnas.org/content/114/12/3017.
4. Daniel Holz, Scott Hughes, and Bernard Schutz, 'Measuring Cosmic Distances with Standard Sirens'(2018) 71(12) *Physics Today* 34, https://physicstoday.scitation.org/doi/full/10.1063/PT.3.4090
5. The Schwarzschild radius of the Sun is about 3 km, so the Schwarzschild radius of a 30 solar mass black hole is about 100 km. Multiplying by 2π gives its circumference as about 600 km. The distance travelled in orbiting 200 times per second is 120,000 km. The

speed of light in a vacuum is 300,000 km per second. So, according to this crude estimate, the black holes were travelling at about 40% of the speed of light.

6. The mass of a hydrogen atom is 939 MeV. The ionization energy of a hydrogen atom is 13.6 eV. The binding energy corresponds to 1.4×10^{-6}% of the mass of the atom.

7. Such as through a photon interacting with the hydrogen atom, in a chemical reaction or by heating hydrogen to a few thousand degrees.

8. The Royal Swedish Academy of Sciences, Scientific Background on the Nobel Prize in Physics 2017, https://www.nobelprize.org/uploads/2018/06/advanced-physicsprize2017-1.pdf.

9. James Joyce's, perhaps unreadable, multilayered epic *Finnegans Wake* takes place over the course of a single day and simultaneously over an entire cycle of the universe. Each of the ten ages of the universe commences with a thunder word, the first of which on page one is the 100-letter word: 'bababadalgharaghtakamminarronnkonnb ronntonnerronntuonnthunntrovarrhounawnskawntoohoohoorden enthurnuk'. This word is formed in typical Joycean fashion by concatenating words meaning *thunder* in various different languages.

10. Black holes do not have a solid surface, so we do not expect to see any electromagnetic signal when two black holes collide. The collision of the black holes' accretion discs might produce an electromagnetic signal, but it would be too faint to see from the vast distances probed by the gravitational wave detectors.

11. The Einstein Telescope website http://www.et-gw.eu/.

12. Cosmic Explorer website https://cosmicexplorer.org/#overview.

Chapter 9

1. Gravity is the only force we know of whose reach extends across intergalactic distances, so it plays a critical role in cosmology, which is the study of the universe in its entirety. (Electromagnetism is also a long-range force, but there are equal quantities of positively charged and negatively charged particles that tend to cancel out each other's effects over long distances.)

2. The expansion of the universe had been anticipated by the Belgian cleric Lemaitre based on his analysis of general relativity when

applied to the entire universe. Lemaitre had even published red shift data a couple of years before Hubble.

3. The correlation between red shift and distance that implies that the universe is steadily expanding is now officially known as the Hubble–Lemaître Law.

4. Just 2.3 arcminutes by 2 arcminutes.

5. Mitchell Begelman and Martin Rees, *Gravity's Fatal Attraction: Black Holes in the Universe*, 2nd edn (Cambridge: Cambridge University Press, 2010) p.111.

6. Begelman and Rees, *Gravity's Fatal Attraction: Black Holes in the Universe*, 2nd edn, p.119.

7. After a star has collapsed, it will be spinning much faster, due to conservation of angular momentum. It is therefore natural to expect that white dwarfs and neutron stars will be spinning very rapidly. This is readily confirmed in the case of neutron stars, because we can detect the pulsars that they generate. As black holes are even smaller than neutron stars, we would expect them to be spinning even faster. Furthermore, the black hole's spin is expected to increase as it accumulates more material from its swirling accretion disc. It is becoming possible to analyse the X-rays from the accretion disc of distant supermassive black holes to determine their rate of spin, and the early evidence suggests that typically they are spinning at close to the maximum rate allowed by general relativity.

8. Begelman and Rees, *Gravity's Fatal Attraction: Black Holes in the Universe*, 2nd edn, p.119.

9. The galaxy appears to be embedded within a sphere of dark matter that has a fairly uniform distribution. Within a uniform spherical mass, the period of the orbits should be the same, irrespective of the radius of the orbit—just like a conical pendulum. There is observational evidence to suggest that the rotation rate is constant in the outer reaches of a galaxy. If this held true right to the centre of the galaxy, then it would mean that the stars at the centre would be moving much more slowly than the stars that are further out.

10. Fulvio Melia, *The Black Hole at the Centre of the Galaxy* (Princeton, NJ: Princeton University Press, 2003) p.40.

11. AM Ghez et al, *Measuring Distance and Properties of the Milky Way's Central Supermassive Black Hole with Stellar Orbits*.

12. The angular resolution of the HST is 0.05 arcseconds. The apparent size of the Schwarzschild apparent size of the radius of the black hole at the centre of the galaxy is 10 micro-arcseconds.

13. http://www.eventhorizontelescope.org/docs/Doeleman_event_horizon_CGT_CFP.pdf

14. https://aasnova.org/2016/03/23/dance-of-two-monster-black-holes/

15. The orbital plane of the secondary black hole also precesses due to the effect of the spin of the primary black hole on the trajectory of the secondary. This is known as the Lense–Thirring effect.

16. https://en.wikipedia.org/wiki/OJ_287

17. This justifies Einstein's assumption that the universe is homogeneous when applying general relativity to the universe as a whole. This assumption is known as the Cosmological Principle.

18. In the first couple of minutes after the Big Bang, the temperature and density of the universe were sufficiently high to generate nuclear fusion reactions that created nuclei of deuterium (heavy hydrogen), helium and traces of other elements. Calculations based on our understanding of the immediate aftermath of the Big Bang match the observed abundances of these isotopes very well. This is one of the great successes of cosmology and is one pillar of evidence for the Big Bang. However, if the density of ordinary matter was much higher in the early universe, this match between observation and theory would be lost.

19. The universe is, of course, not flat like a pancake or the galaxy. The three-dimensional space that forms the universe is flat in the sense that, on the largest scale, its geometry is best described by Euclidean geometry. Clusters of galaxies form gravitational lenses, but these are like little dimples within what is essentially flat space.

20. Dante, *The Divine Comedy 3: Paradise* (trans. Dorothy L Sayers) (London: Penguin, 1962). In this canto Dante reaches the primum mobile and compares the vast hierarchy of the angelic host that he sees to the grains of wheat on a chess board, as set out in the doubling problem.

21. Ibn Khallikan's *Biographical Dictionary* Vol.3 (1814), p.71 ffIt is available in the following modern edition: Ibn Khallikan's *Biographical Dictionary*, Vol. III. Cosimo Classics (New York, 2013).

22. The exact total of grains of wheat on the chess board can be calculated as follows. If this total is called S, then

$$S = 1 + 2 + 4 + \ldots + 2^{62} + 2^{63}.$$

We can multiply both sides of this sum by 2, to give:

$$S = 2 + 4 + 8 + \ldots + 2^{63} + 2^{64}.$$

Now, all the terms in these two sums are identical except the first term in the first sum and the last term in the second sum. If we subtract the sum corresponding to S from the sum corresponding to 2S, this gives:

$$2S - S = 2^{64} - 1.$$

Therefore, the total number of grains of wheat on the chess board is: $S = 2^{64} - 1$.

23. Steven Weinberg, *Cosmology* (Oxford: Oxford University Press, 2008) p.212.

Image Credits

Chapter 2 includes material originally written for an article by Nicholas Mee published as Venus in the Face of the Sun in the June 2012 issue of History Today.

Index